CHEMISTRY FOR
AGRICULTURE AND ECOLOGY

TO THE STAFF AND
STUDENTS AT BUNDA
1974–1976

CHEMISTRY FOR AGRICULTURE AND ECOLOGY

A FOUNDATION COURSE

R. K. M. HAY

BSc, MSc, PhD
Department of Environmental Sciences
University of Lancaster

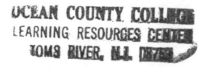
BLACKWELL SCIENTIFIC PUBLICATIONS
OXFORD LONDON EDINBURGH
BOSTON MELBOURNE

© 1981 by
Blackwell Scientific Publications
Editorial offices:
Osney Mead, Oxford, OX2 0EL
8 John Street, London, WC1N 2ES
9 Forrest Road, Edinburgh, EH1 2QH
52 Beacon Street, Boston
     Massachusetts 02108, USA
214 Berkeley Street, Carlton
     Victoria 3053, Australia

First published 1981

Typeset by Enset Ltd
Midsomer Norton, Bath, Avon
and printed and bound
in Great Britain by
Billing & Sons Ltd
Guildford, London, Oxford, Worcester

DISTRIBUTORS

USA
  Blackwell Mosby Book Distributors
  11830 Westline Industrial Drive
  St Louis, Missouri 63141
Canada
  Blackwell Mosby Book Distributors
  120 Melford Drive, Scarborough
  Ontario, M1B 2X4
Australia
  Blackwell Scientific Book
  Distributors
  214 Berkeley Street, Carlton
  Victoria 3053

British Library
Cataloguing in Publication Data

Hay, Robert K. M.
  Chemistry for agriculture and
     ecology.
  1. Chemistry
  2. Agricultural chemistry
  3. Ecology
  I. Title
  540′.2′463   QD31.2

  ISBN 0-632-00699-4

CONTENTS

v

PREFACE

An understanding of the basic principles of chemistry is essential in the study of almost all branches of biology and applied biology. In spite of this, there is a large, and perhaps increasing, number of students embarking on university or college courses in agriculture and ecology with only the sketchiest knowledge of the properties and reactions of natural materials. In many cases this deficiency can be a serious handicap throughout their course of study.

This book was written primarily for these students. Assuming no prior knowledge of chemistry and only an elementary grasp of the laws of physics and algebra, it covers the fundamental ideas of atomic, physical and organic chemistry using examples drawn largely from the environmental sciences. Because of the author's teaching and research interests, these examples are biased somewhat towards the plant and soil sciences; however, this should be of positive benefit to agriculturalists and ecologists since almost all other books covering the 'Chemistry for Biology' field are really elementary texts in biochemistry, written for students of medicine and microbiology. Inevitably, in a book of this kind, there are sections which are not of universal interest. For example, ecologists will be less interested in the structure and properties of metals than agriculturalists, who should find the information useful in subsequent studies in agricultural engineering. On the other hand, the problems of water quality and purification may be of greater interest to the ecologist. However, most of the book should be of value to both groups, sometimes for different reasons; for example, in Part 3, agricultural chemicals (fertilizers and pesticides) are considered both as indispensable tools in agricultural production and as potential environmental pollutants.

However, it cannot be stressed sufficiently that this is a *foundation* course and that students embarking on more advanced courses in, say, environmental chemistry, soil science or biochemistry, will need to make full use of the reading lists provided at the end of each chapter. In particular, some students will need to build a more rigorous understanding of topics such as thermodynamics (including

ix

redox potentials), chemical kinetics, radioisotope chemistry and physics, and organic reaction mechanisms, on the foundation laid by the contents of this book. It is hoped that the unusual importance given to agricultural and ecological examples, together with the recommended reading lists will make this book useful to a wide range of students of varying chemical competence.

This book began life in 1975 as a duplicated 'Chemistry Handbook' issued to my students at Bunda College, University of Malawi. In a revised form, it has been used since 1978 as the standard first-year chemistry text by diploma students at the Lancashire College of Agriculture under the direction of Mike Harbridge, whose help and advice have been invaluable during the final revision. Over the last three years it has also been used by students entering Lancaster University to study ecology, who have difficulties with chemistry.

Of the many colleagues who have read the various drafts of this book, I am particularly grateful for the substantial critical contributions of Professor J.A. Leiston, Dr M.J. Jones and Miss Cerys White (University of Malawi), Dr K. Smith and Dr Carol Duffus (Edinburgh School of Agriculture) and Professor A. Wild (University of Reading). I should also like to thank Carolyn Amos, Jackie Dixon and Paul Taylor for their great help in preparing the manuscript. The preparation and production of this book have been very much facilitated by the helpful approach of Robert Campbell and his assistants at Blackwell Scientific Publications. Finally, it is clear that a book such as this can be of value only if it meets the needs of the students for which it is written. I should, therefore, be pleased to receive any critical comments, or suggestions for improvement of the text.

Robert K.M. Hay

General note. One universal scientific concept, which is not discussed specifically in the text, is the idea of a *Law* or *Principle*, meaning a generalization about the behaviour of matter (in this case chemical substances) formulated from the results of many *experiments* (e.g. the Gas Laws, section 6.1; Law of Mass Action and Le Chatelier's Principle, section 7.6). They are, therefore, ways of summarizing the results of past experiments and predicting the outcome of future experiments (contrast with legal laws which are *imposed*). In the *unlikely* event of a law not being obeyed in an experiment, the law must be reformulated to accommodate the unexpected results.

ACKNOWLEDGEMENTS

The author is grateful to the following authorities for permission to reproduce material from their publications:

Cambridge University Press (Figs 4.1, 4.2 & 4.3); Professor J. Zussman and Longmans (Fig. 4.6); W.H. Freeman & Co. (Table 4.4); Dr. R. Scott Russell and McGraw-Hill Book Co. (UK) Ltd. (Table 5.3); Dr. M. Slessor and Blackie & Son Ltd. (Table 7.1); Macmillan Publishing Co. Inc. (Fig. 8.1; Table 10.4); Professor H.A. Scheraga and American Institute of Physics (Fig. 9.1); Van Nostrand Reinhold Co. and Prentice-Hall Inc. (Table 14.3); Dr. J.C. Dearden and New Science Publications (Fig. 14.1); Professor J.C. Kendrew and Macmillan Journals Ltd. (Fig. 14.2); Macmillan Publishing Co. Ltd. (Fig. 14.3).

The author also wishes to acknowledge the great usefulness of the Handbook of Chemistry and Physics (The Chemical Rubber Co.) as a source of chemical data.

PART 1
ATOMIC AND PHYSICAL
CHEMISTRY

CHAPTER 1
ATOMIC THEORY

1.1 ATOMS

If we take a piece of pure iron and divide it up continuously, there will come a time when our tools, however sharp, will not be able to cut the fragments into smaller pieces. However, if we imagine that we have a tool with which we can subdivide the metal much further, the question eventually arises 'Is the metal made up of identical particles which cannot be cut up into yet smaller identical particles?'.

Atomic theory gives the answer 'yes' to this question, and according to the theory the smallest identical particles are called atoms. For example, it can be calculated that a 10 g sample of pure iron contains $1 \cdot 1 \times 10^{23}$ iron atoms, each atom being $2 \cdot 5 \times 10^{-10}$m in diameter. The same is true of all materials in the world; for example, the pens we write with, the clothes we wear, the air we breathe and even our bodies, are made up of many millions of atoms held together by chemical bonds.

These tiny atoms are not all alike. The atoms of each element have the same mass (weight), whereas atoms of different elements have different masses. (Over one hundred different elements have been discovered, e.g. hydrogen, sodium, iron, uranium, see Table 1.2.) Our piece of iron contains only one kind of atom, the iron atom, but this is unusual. Pure elements are not normally encountered outside chemical laboratories and most substances we use are compounds, made up of a variety of kinds of atom firmly bonded together. For example, water is a compound containing two different types of atom (hydrogen and oxygen) whereas 'Maneb' (a widely used fungicide) contains five different kinds of atom (carbon, hydrogen, nitrogen, sulphur and manganese).

1.2 SUB-ATOMIC PARTICLES

Although atoms are the smallest *identical* particles in our lump of iron, each atom can be subdivided into a nucleus, which is positively

TABLE 1.1. Number of protons, neutrons and electrons in the atoms of a selection of elements.

Element	Protons	Neutrons	Electrons
Hydrogen	1	0	1
Helium	2	2	2
Nitrogen	7	7	7
Phosphorus	15	16	15
Potassium	19	20	19

charged, and a number of electrons, which are negatively charged. In turn, the nucleus can be divided up into a number of positive protons and a number of uncharged neutrons. Since atoms are not electrically charged, the number of electrons in an atom is always the same as the number of protons, as shown, for some small atoms, in Table 1.1.

Almost all of the mass of an atom is in the nucleus. For example, in the simplest atom (hydrogen) the nucleus (1 proton) accounts for 99·95% of the total mass, the single electron contributing only 0·05%. On the other hand, the volume of the nucleus is tiny

TABLE 1.2. Atomic properties of the elements.

Element	Atomic number	Atomic mass	Element	Atomic number	Atomic mass
Hydrogen(H)	1	1·008(1)	Potassium(K)	19	39·10(39)
Helium(He)	2	4·003(4)	Calcium(Ca)	20	40·08(40)
Lithium(Li)	3	6·94(7)	Scandium(Sc)	21	44·96(45)
Beryllium(Be)	4	9·01(9)	Titanium(Ti)	22	47·90(48)
Boron(B)	5	10·81(11)	Vanadium(V)	23	50·94(51)
Carbon(C)	6	12·01(12)	Chromium(Cr)	24	52·00(52)
Nitrogen(N)	7	14·007(14)	Manganese(Mn)	25	54·94(55)
Oxygen(O)	8	16·00(16)	Iron(Fe)	26	55·85(56)
Fluorine(F)	9	19·00(19)	Cobalt(Co)	27	58·93(59)
Neon(Ne)	10	20·18(20)	Nickel(Ni)	28	58·71(59)
Sodium(Na)	11	23·00(23)	Copper(Cu)	29	63·54(63·5)
Magnesium(Mg)	12	24·31(24)	Zinc(Zn)	30	65·37(65)
Aluminium(Al)	13	26·98(27)	Gallium(Ga)	31	69·72(70)
Silicon(Si)	14	28·09(28)	Germanium(Ge)	32	72·59(73)
Phosphorus(P)	15	30·97(31)	Arsenic(As)	33	74·92(75)
Sulphur(S)	16	32·06(32)	Selenium(Se)	34	78·96(79)
Chlorine(Cl)	17	35·45(35·5)	Bromine(Br)	35	79·91(80)
Argon(Ar)	18	39·95(40)	Krypton(Kr)	36	83·80(84)

compared with that of the whole atom. There are many ways of illustrating this; for example, if the nucleus of a hydrogen atom were enlarged so as to have the diameter of a pinhead (about 1 mm), then the diameter of the whole atom would be about 100 metres. Thus we should imagine the nucleus of an atom as a minute, extremely dense, body inside a large sphere, which is nearly empty apart from a few rapidly circulating electrons.

At this point it should be stressed that an electron from a hydrogen atom is identical to an electron from any other element. The same is true for protons and neutrons. Atoms of one element are distinguished from those of another element by the number of protons in the nucleus (the Atomic Number), and not by any difference in the protons (electrons or neutrons) themselves.

1.3 ATOMIC MASS AND ISOTOPES

For each element we have a characteristic Atomic Mass (Table 1.2). The mass of one proton is almost identical to the mass of one neutron and this mass is defined as 1 dalton (after Dalton who first formulated the Atomic Theory). Since almost all of the mass of an atom is in the nucleus, the atomic mass is the sum of the numbers of protons and neutrons in the nucleus, expressed in daltons. For example, the phosphorus atom has 15 protons and 16 neutrons in the nucleus, giving an atomic mass of $15+16 = 31$ daltons. (In future, we shall omit the units of atomic mass.)

However, it is clear from Table 1.2 that elements do not have whole number atomic masses as we would expect from the method of calculation. The reason is that most elements have more than one isotope. Isotopes are atoms which have the same atomic number but different atomic masses, i.e. they have the same number of protons and electrons but different numbers of neutrons. For example, as shown in Table 1.3, there are three isotopes of hydrogen, of which the protium isotope is by far the most common, with deuterium occurring naturally to the extent of only 1 atom per 6000 atoms of protium. Tritium is a short-lived radioactive isotope (see section 1.7) which does not occur naturally but can be synthesized in radiochemical laboratories. Due to the preponderance of the protium isotope, the atomic mass of hydrogen is nearly 1 ($1\cdot008$).

In the case of chlorine, we have an atomic mass of $35\cdot45$. This value arises because about three-quarters of the atoms in a sample of

TABLE 1.3. Properties of the isotopes of hydrogen.

Isotope	Number of protons	Number of electrons	Number of neutrons	Atomic number	Atomic mass
Protium	1	1	0	1	1
Deuterium (D)	1	1	1	1	2
Tritium (T)	1	1	2	1	3

chlorine gas have the $^{35}_{17}Cl$ isotope structure (17 protons, 18 neutrons, atomic mass 35), whereas the remaining quarter have the $^{37}_{17}Cl$ isotope structure (17 protons, 20 neutrons, atomic mass 37).

Although Table 1.2 shows precise atomic mass values, whole number values (1 for H, 23 for Na, 31 for P, etc.) are sufficiently accurate for chemical calculations, except in the cases of chlorine (35·5) and copper (63·5).

Having considered isotopes, we are now in a position to give an exact definition of an element, i.e. *an element* is a substance composed of atoms of the same atomic number (number of protons in the nucleus). This is a most useful definition since it stresses that the chemical differences between elements (properties and reactions) arise from differences in numbers of protons and electrons but not neutrons. As we shall see in the next chapter, the chemical reactivity of an element (the ability to form chemical bonds) depends on the number and arrangement of *electrons* in the atoms of the element. The number of neutrons is important only in determining physical properties, especially atomic mass.

1.4 THE PERIODIC TABLE, PART 1

In nature, there are more than a hundred elements and many millions of compounds of these. Before we can consider a systematic study of this bewildering array of chemical substances, we must first classify them to stress the similarities between closely related elements, and the differences between more distantly related elements.

We can begin to classify the elements by writing them out in order of atomic number and looking for similarities between them:

i.e.　H, He, Li, Be, B, C, N, O, F, Ne, Na, Mg, Al, Si, P, S, Cl, Ar, K, Ca – – – – – – – – – – – – –.

For example, it can be demonstrated in the laboratory that Li, Na and K are very similar elements. They are all extremely reactive and soft metals. On the other hand, He, Ne and Ar are all extremely un-reactive gases. Similar relationships exist between Be, Mg and Ca; B and Al; C and Si; N and P; O and S, and F and Cl. Using this in-formation, it is possible to start constructing a table of elements by writing closely related elements in vertical columns:

							He
Li	Be	B	C	N	O	F	Ne
Na	Mg	Al	Si	P	S	Cl	Ar
K	Ca						

Beyond Ca, difficulties begin to arise. The next inert gas like Ar is Kr but there are fifteen elements with atomic masses falling between Ca(40) and Kr(84) and not all of these elements can fit into the five vacant spaces in our table. However, ten of these elements (Sc, Ti, V, Cr, Mn, Fe, Co, Ni, Cu, Zn) are very similar to one another in properties and can be grouped together as 'transition elements'. The table now becomes:

H

																	He
Li	Be											B	C	N	O	F	Ne
Na	Mg											Al	Si	P	S	Cl	Ar
K	Ca	Sc	Ti	V	Cr	Mn	Fe	Co	Ni	Cu	Zn	Ga	Ge	As	Se	Br	Kr

The remaining elements can be fitted into the table in a similar way, giving the full Periodic Table (Table 1.4) in which the vertical columns (containing closely related elements) are called Groups. Li is the lightest element in Group 1, Be in Group 2, B in Group 3 and so on. The transition metals are grouped separately. The horizontal rows are termed Periods; period 1 contains H and He, period 2 Li to Ne and so on. It is difficult to find a place for H in the periodic table since it has similarities to groups 1 and 7. The simplest solution is to place it alone above the table. At the bottom of the table, the arrangement is complex due to the Lanthanide and Actinide elements which are of little interest to us.

1.5 SHELL FILLING—THE ARRANGEMENT OF ELECTRONS IN ATOMS

In the periodic table, the most unreactive elements are in group 8, the inert gases, which will form compounds with other elements only under extreme conditions. As we have noted earlier, chemical

TABLE 1.4. The periodic table of the elements.

1 ← Atomic Number
H
1·008 ← Atomic Mass

1	2	3	4	5	6	7	8	9	10	11	12	13	14	15	16	17	18
1 H 1·008																	2 He 4·00
3 Li 6·94	4 Be 9·01											5 B 10·81	6 C 12·01	7 N 14·01	8 O 16·00	9 F 19·00	10 Ne 20·18
11 Na 23·0	12 Mg 24·3											13 Al 27·0	14 Si 28·1	15 P 31·0	16 S 32·1	17 Cl 35·5	18 Ar 39·9
19 K 39·1	20 Ca 40·1	21 Sc 45·0	22 Ti 47·9	23 V 50·9	24 Cr 52·0	25 Mn 54·9	26 Fe 55·9	27 Co 58·9	28 Ni 58·7	29 Cu 63·5	30 Zn 65·4	31 Ga 69·7	32 Ge 72·6	33 As 74·9	34 Se 79·0	35 Br 79·9	36 Kr 83·8
37 Rb 85·5	38 Sr 87·6	39 Y 88·9	40 Zr 91·2	41 Nb 92·9	42 Mo 95·9	43 Tc 99	44 Ru 101·1	45 Rh 102·9	46 Pd 106·4	47 Ag 107·9	48 Cd 112·4	49 In 114·8	50 Sn 118·7	51 Sb 121·8	52 Te 127·6	53 I 126·9	54 Xe 131·3
55 Cs 132·9	56 Ba 137·3	57–71 SEE BELOW	72 Hf 178·5	73 Ta 180·9	74 W 183·9	75 Re 186·2	76 Os 190·2	77 Ir 192·2	78 Pt 195·1	79 Au 197·0	80 Hg 200·6	81 Tl 204·4	82 Pb 207·2	83 Bi 209·0	84 Po 209	85 At 210	86 Rn 222
87 Fr 223	88 Ra 226	89– SEE BELOW															

57 La 138·9	58 Ce 140·1	59 Pr 140·9	60 Nd 144·2	61 Pm 147	62 Sm 150·4	63 Eu 152·0	64 Gd 157·3	65 Tb 158·9	66 Dy 162·5	67 Ho 164·9	68 Er 167·3	69 Tm 168·9	70 Yb 173·0	71 Lu 175·0
89 Ac 227	90 Th 232·0	91 Pa 231	92 U 238·0	93 Np 237	94 Pu 242	95 Am 243	96 Cm 247	97 Bk 245	98 Cf 251	99 Es 254	100 Fm 253	101 Md 256	102 No 254	103 Lw 257

Key to common elements:

Rb	Rubidium	Sr	Strontium	Mo	Molybdenum	Ag	Silver	Cd	Cadmium	Sn	Tin	I	Iodine
Cs	Caesium	Ba	Barium	W	Tungsten	Pt	Platinum	Au	Gold	Hg	Mercury	Pb	Lead

reactivity depends upon the electrons and so we must assume that the 2 electrons in He, 10 in Ne, 18 in Ar, 36 in Kr, 54 in Xe and 86 in Rn must be arranged in a special way which makes these elements resistant to chemical attack. We say that the Inert Gases have a stable arrangement of electrons and, in general, they cannot achieve greater stability by reaction with other elements.

This statement brings us to another part of the atomic theory—the filling of shells. We must imagine that the atom is like the solar system with the nucleus at the centre instead of the sun. The electrons are like planets circling the nucleus in orbits, orbitals or shells. These shells can contain only a certain number of electrons.

We can illustrate shell filling by considering the elements one by one. For example, in Hydrogen (1 proton, 0 neutrons, 1 electron), the single electron circles the nucleus in the first (K) shell. In Helium (2 protons, 2 neutrons, 2 electrons), the two electrons occupy the same K shell—

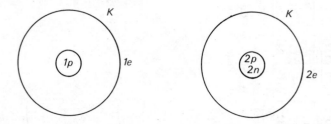

However, since we know that He is very unreactive and stable, we *assume* that this first (K) shell is now full. Consequently, when we come to Li (3 protons, 4 neutrons, 3 electrons), the 3rd electron cannot be accommodated in the first shell and must be placed in the second (L) shell—

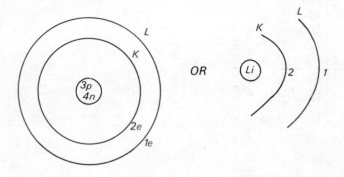

The elements succeeding Li are then as shown—

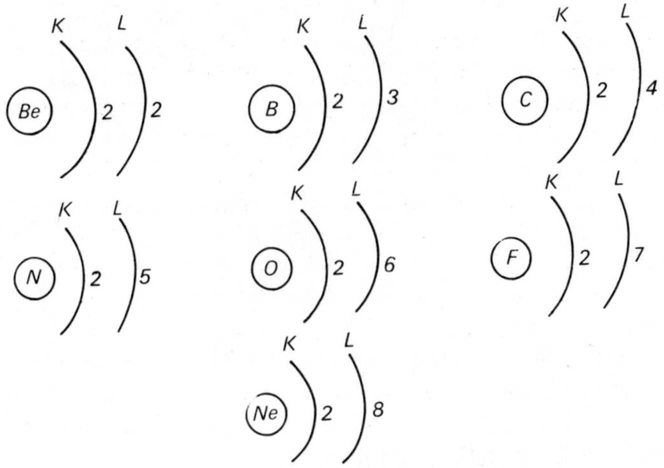

Since Ne is an inert gas with a stable electronic structure, we conclude that the second (L) shell, containing 8 electrons, is now full, or closed. Thus the L shell can hold a maximum of 8 electrons. Additional electrons must go into a third (M) shell—

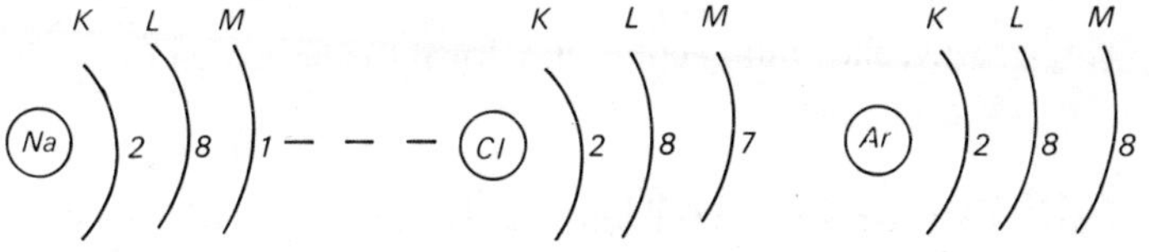

We can now see why Na and Li behave similarly in chemical reactions—each has a single electron in the outermost shell. The electrons accommodated in inner, closed, shells do not normally take part in reactions, and it is only the electrons in the outermost shell which determine the reactivity of the element. (This will become clearer when we consider bonding in Chapter 2.) Thus F and Cl which each have seven electrons in the outer shell, have very similar chemical properties.

At Argon, we have again reached a stable state and the shell will be closed. However, we have a complication here. Shell 3 (M) can contain a maximum of 8 electrons in Ar, K and Ca only. When we reach the transition elements, electrons again begin to enter the M shell which can now contain up to 18 electrons. This can be illustrated as follows—

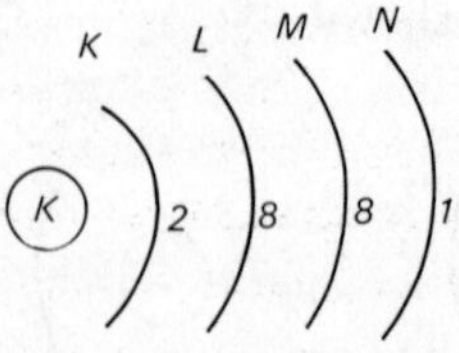

(*Compare with Li & Na*) 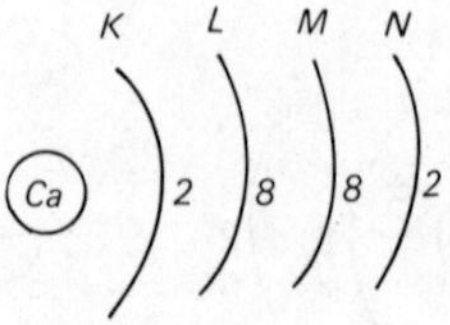   (*Compare with Be & Mg*)

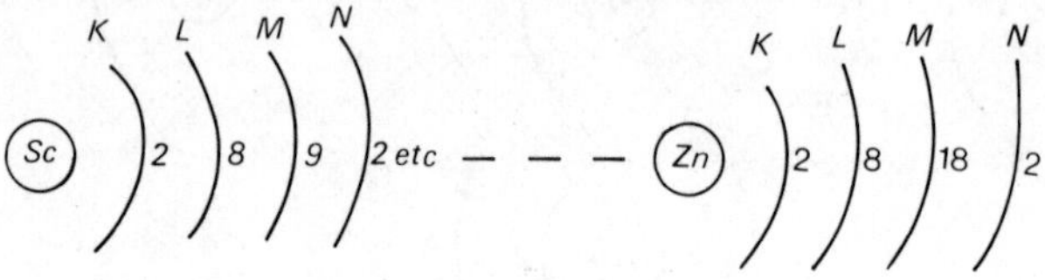

The N shell now fills up normally—

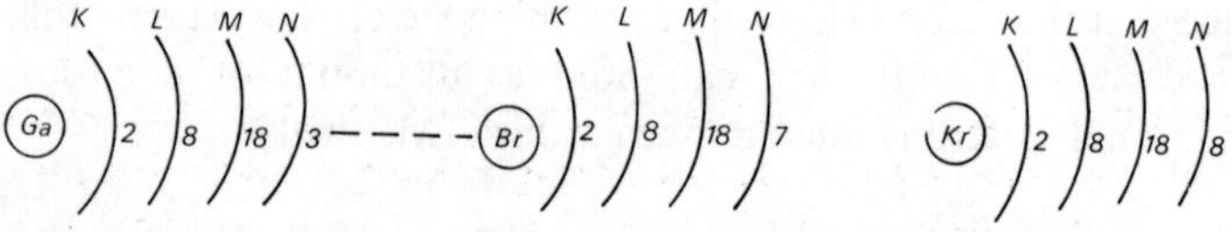

In a similar way, shell filling continues down the periodic table.

1.6 THE PERIODIC TABLE, PART 2

We have seen why the periodic table is a table of chemical reactivity. Elements with 1 electron in the outermost shell are placed in group 1, those with 2 electrons in group 2 and so on. As well as placing elements in their correct order of atomic number, the periodic table also arranges elements in a definite pattern in relation to electronegativity and electropositivity.

The *Electronegativity* of an element is an index of the ability of atoms of the element to *attract* electrons. The *Electropositivity* of an element is an index of the extent to which atoms of the element will tend to *lose* electrons. (Obviously an element with a high electronegativity will have a low electropositivity, and vice versa.)

As we move from the left side of the table to the right, e.g. Na to Cl, the elements in any period become progressively more electronegative (and less electropositive), i.e. Na is highly electropositive, so much so that it tends to lose its outer electron when it enters into a compound. At the other extreme, the Cl atom which is highly electronegative, tends to hold its electrons firmly, and attempts to gain another electron when it forms compounds. Electronegativity also decreases as we go down each group and consequently, electropositivity increases down the table. Thus we have the situation where Cs is more electropositive than Li whereas F is more electronegative than I. In general, elements in the top right-hand side of the periodic table are highly electronegative, whereas elements in the bottom left-hand side are highly electropositive. The remaining elements are intermediate (Table 1.5).

TABLE 1.5. Electronegativities of selected elements.

			H			
			2·1			
Li	Be	B	C	N	O	F
1·0	1·5	2·0	2·5	3·0	3·5	4·0
Na	Mg	Al	Si	P	S	Cl
0·9	1·2	1·5	1·8	2·1	2·5	3·0
K	Ca		Ge	As	Se	Br
0·8	1·0		1·8	2·0	2·4	2·8
Rb	Sr		Sn	Sb	Te	I
0·8	1·0		1·8	1·9	2·1	2·5
Cs	Ba		Pb	Bi		
0·7	0·9		1·8	1·9		

The characteristic properties of metals depend upon the ability of the atoms of the metal to lose electrons, i.e. their electropositivity. For example, the fact that metals conduct electricity (electrons) well is a result of the free movement of outer shell electrons. Since electropositivity increases from right to left and from top to bottom of the periodic table, it follows that in each succeeding period, more elements will have metallic properties. In period 2, only Li and Be are metals; in period 3, Na, Mg and Al are metals; in period 4, Ge in group 4 is a metal in addition to the 13 preceding elements. In period 5, metallic properties spread to group 5 and to group 6 in period 6. Overall, the majority of elements are metallic.

These ideas of electronegativity and electropositivity are essential for the next chapter where there is a discussion of the bonding of atoms.

1.7 RADIOACTIVE ISOTOPES

In section 1.3, we noted that, of the three isotopes of hydrogen which can exist, tritium is unstable and must be synthesized artificially. In the same way, $^{12}_{6}C$ is stable whereas the nuclei of $^{14}_{6}C$ tend to disintegrate. At least one unstable, or radioactive, isotope exists for each of the elements.

The nuclei of radioactive isotopes break down at different rates. For example, half of a sample of tritium will disintegrate in 12·3 years (one half-life) compared with 12·4 hours for $^{42}_{19}K$ and 5730 years for $^{14}_{6}C$. One quarter of the original sample will remain after a further half-life, and so on. Consequently, of the three isotopes mentioned, only $^{14}_{6}C$ occurs naturally.

When they disintegrate, atoms of radioactive isotopes release energy in the form of α, β or γ rays which can be detected and measured by an appropriate counter. Because of this, radioactive isotopes, used as *tracers*, are probably the most revolutionary research tools in ecology and agriculture.

For example, we can expose an actively photosynthesizing leaf to 'labelled' carbon dioxide, in which a small proportion of the carbon atoms are $^{14}_{6}C$ instead of $^{12}_{6}C$. As absorption and assimilation of carbon dioxide proceed, tracer carbon atoms (which to the plant are indistinguishable from $^{12}_{6}C$ atoms) become incorporated into organic molecules and it is possible to reconstruct the series of biochemical reactions of photosynthesis by counting the $^{14}_{6}C$ content of a variety of compounds at different times. Over longer periods of time, the translocation of photosynthate from the site of synthesis in the leaf to the site of use or storage in the plant can be charted by monitoring the radiocarbon content of different plant tissues.

Tracers are employed in a wide range of environmental investigations; for example in studies of:

(i) the movement of marine and freshwater sediments;

(ii) the fate of fertilizers, pesticides and pollutants in soils, plants and water;

(iii) the passage of substances (nutrients, drugs, etc.) through large animals (avoiding the wasteful slaughter of animals involved in destructive sampling);

(iv) the age of organic materials (e.g. peat deposits); this application is based on the fact that the natural $^{14}_{6}C$ content of the peat material at the time of deposition falls in a predictable way, according to the half-life, i.e. 50% lost after 5730 years, 75% after 11 460 years, 87·5% after 17 190 years, etc.

With the development of nuclear power stations, radioactive isotopes are also becoming important as environmental pollutants.

In experimental work, synthetic radioactive isotopes, prepared in radiochemical laboratories, are normally used. These laboratories offer a wide range of short- and long-lived isotopes, as well as many organic compounds labelled with $^{14}_{6}C$ and tritium for biological and biochemical studies.

EXERCISES

(1) Using Table 1.4, calculate the numbers of protons, neutrons and electrons in the predominant naturally occurring isotope of the following elements:

B, O, Mg, Si, S, Ca, Co, Zn, As, I.

(2) Draw electronic structure diagrams for all the elements from H to Sr.

FURTHER READING

The ground covered in this chapter, which is common to all studies of elementary chemistry, can be discussed in a number of different ways, with stress laid on different aspects. The following list of suggested reading exemplifies this diversity of approach with 4, for example, laying considerable stress on the *shapes*, as well as the limited accommodation, of electron orbitals

1. DAVIES L. *et al.* (1973) *Investigating Chemistry*. Chs 1 and 5. Heinemann, London.
2. HOFFMAN K.B. (1963) *Chemistry for the Applied Sciences*. Chs 1 and 2. Prentice-Hall, Inc., New Jersey.
3. ROSSOTTI H. (1975) *Introducing Chemistry*, Part II. Penguin Books, Middlesex.
4. WHITE E.H. (1970) *Chemical Background for the Biological Sciences*. 2nd edition, Ch. 1. Prentice-Hall, Inc., New Jersey.

Note also:
5. HARPER D. (1963) *Isotopes in Action*. Pergamon Press, Oxford. (Discusses a wide range of applications of radioactive isotopes.)
6. *Handbook of Chemistry and Physics* (various editions). The Chemical Rubber Company. (The authoritative compendium of chemical information.)

CHAPTER 2
CHEMICAL BONDS

In our discussion of atomic theory in Chapter 1, we concentrated on elements. Some familiar materials, such as aluminium and sulphur, are elements but most of the substances we encounter in agriculture and ecology are compounds, i.e. substances which contain atoms of two or more elements combined in a definite numerical ratio. For example:

(i) *Ammonium Sulphate*, a widely used fertilizer, contains atoms of the elements N, H, S and O combined in the ratio 2 : 8 : 1 : 4.

(ii) *Uric Acid*, an important excretory product of terrestrial insects, contains C, O, N and H atoms combined in the ratio 5 : 3 : 4 : 4.

How are the atoms held together in these definite ratios?

In our discussion of the Periodic Table, we saw that the arrangement of electrons in atoms of the inert gases was more stable than in the other elements. The atoms of the less stable (and more reactive) elements can also achieve the same stable arrangement of electrons as exist in the inert gases, by exchanging or sharing electrons with other atoms. This exchange or sharing of electrons results in the formation of chemical bonds which bind the atoms together.

2.1 IONIC OR ELECTROVALENT BONDS

Ionic bonds are formed by the *exchange* of electrons between atoms. For example, we can explain how the K and Cl atoms in potassium chloride (a potassium fertilizer, used under the archaic name muriate of potash) are held together, in the following way.

As we saw on page 12, the K atom has one electron in its outer shell. In order to achieve the stability of an 'inert gas structure', each K atom needs to lose one electron, thereby forming a charged atom (or ion) K^+, with the electronic structure of Ar. On the other hand, Cl contains seven electrons in its outer shell. To gain an 'inert gas structure', this atom has to gain an additional electron. This will result in a charged ion, Cl^-, which also has the electronic structure of Ar.

It will now be clear that if one electron is transferred from a K atom to a Cl atom, then each atom will achieve a stable electronic state similar to Ar. As a result of this electron exchange, both atoms become charged and the positive K ion (cation) will attract the negative Cl ion (anion). This attraction between opposite charges constitutes an ionic bond.

The simplest way to illustrate ionic bonds is by 'dot diagrams' where the outermost shell electrons are indicated by dots (or crosses) as follows—

$$\overset{\circ}{K} + \overset{\bullet\bullet}{\underset{\bullet\bullet}{\bullet Cl \bullet}} \longrightarrow \overset{+}{K} + \overset{\bullet\bullet}{\underset{\bullet\bullet}{\circ Cl \bullet}}^{-}$$

Note that the electrons from K and Cl are indistinguishable; different symbols are used simply to stress the origins of the electrons.

2.2 COVALENT BONDS

Covalent bonds are formed by the *sharing* of electrons between atoms. For example, the gaseous element chlorine is made up of pairs of chlorine atoms held together by covalent bonds as described below.

As we saw in section 2.1, each Cl atom needs to gain one electron to achieve the stable electronic configuration of Ar. However, in chlorine, instead of gaining an electron outright as in the ionic bond of KCl, each atom shares one of its seven electrons with another Cl atom, i.e. using a dot diagram—

$$\overset{\circ\circ}{\underset{\circ\circ}{\circ Cl \circ}} + \overset{\times\times}{\underset{\times\times}{\times Cl \times}} \longrightarrow \overset{\circ\circ}{\underset{\circ\circ}{\circ Cl}} \overset{\times\times}{\underset{\times\times}{Cl \times}} \text{ or } Cl - Cl$$

Thus each atom has a 'share' in eight outer electrons (an octet). The shared electrons hold the two atoms together in a single covalent bond giving a diatomic molecule.

In certain circumstances, multiple sharing of electrons can occur, resulting in the formation of double or triple covalent bonds, e.g. O_2 and N_2—

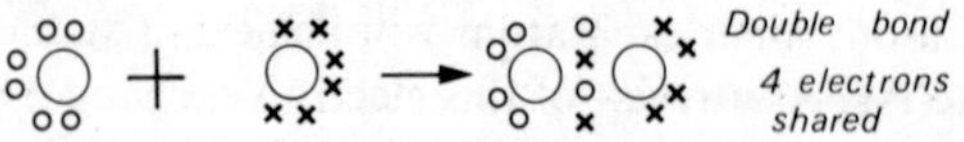

$$\text{:Ö:} + \text{:Ö:} \longrightarrow \text{:Ö::Ö:} \qquad \begin{array}{l} \textit{Double bond} \\ \textit{4 electrons} \\ \textit{shared} \end{array}$$

$$\textit{or} \quad \text{O}=\text{O}$$

$$\text{:N:} + \text{:N:} \longrightarrow \text{:N:::N:} \qquad \begin{array}{l} \textit{Triple bond} \\ \textit{6 electrons} \\ \textit{shared} \end{array}$$

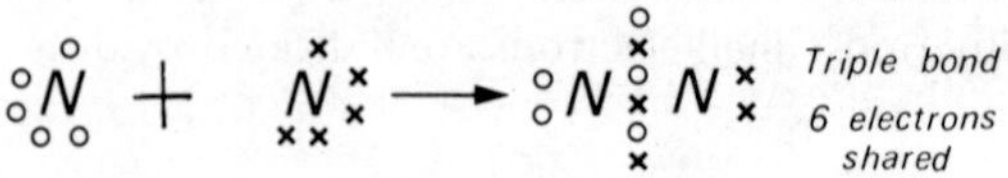

$$\textit{or} \quad \text{N} \equiv \text{N}$$

2.3 SOME COMMON IONIC AND COVALENT COMPOUNDS

As we saw in Table 1.5, atoms of the elements on the left of the Periodic Table are highly electropositive and therefore tend to form cations, by the loss of one electron (e.g. Na^+, K^+), two electrons (Mg^{2+}, Ca^{2+}) or three electrons (Al^{3+}). On the other hand, atoms of the elements in Group 7 (the Halogens) are highly electronegative and tend to give anions (e.g. F^-, Cl^-, Br^-). Consequently, compounds of metals and halogens (the metal halides) are predominantly ionic (e.g. common salt or sodium chloride (NaCl), potassium chloride (KCl) and a number of less familiar substances such as sodium fluoride (NaF), potassium bromide (KBr) and magnesium chloride ($MgCl_2$)).

In these simple ionic compounds, both cations and anions are charged atoms. However, many important ionic substances are made up of polyatomic ions, which consist of a number of atoms covalently bonded together. For example, in sodium hydroxide, NaOH, the oxygen atom achieves a stable octet of electrons by sharing a pair of electrons with a hydrogen atom (covalent bond) and accepting one electron from a sodium atom (ionic bond)—

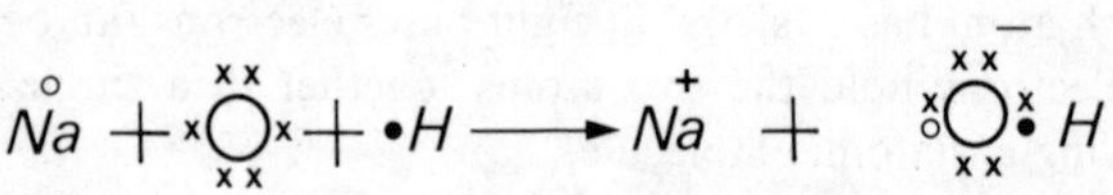

$$Na + \text{:O:} + \cdot H \longrightarrow Na^+ + \text{:Ö:H}^-$$

Thus sodium hydroxide is made up of sodium and hydroxide ions held together by ionic bonds. Since the oxygen and hydrogen atoms are firmly bonded together, the hydroxyl ion does not normally break down in chemical reactions. Consequently, sodium hydroxide is classed as an ionic compound.

Other polyatomic anions, including nitrate (NO_3^-), sulphate (SO_4^{2-}), carbonate (CO_3^{2-}) and different forms of phosphate ($H_2PO_4^-$, HPO_4^{2-} and PO_4^{3-}) occur in ionic substances which are important in agriculture, e.g.

Liming compounds — calcium hydroxide, $Ca(OH)_2$
calcium carbonate, $CaCO_3$
Fertilizers — sodium nitrate, $NaNO_3$
monocalcium phosphate, $Ca(H_2PO_4)_2$.

There is also one important polyatomic cation, the ammonium ion (NH_4^+), which is the primary product of the mineralization of organic nitrogen in soils.

Many important compounds are covalently bonded, e.g.

atmospheric gases — oxygen (O_2), nitrogen (N_2), carbon dioxide (CO_2)
water — (H_2O)
hydrocarbon fuels — methane (CH_4), butane (C_4H_{10})
anaesthetics — chloroform ($CHCl_3$), ethyl chloride (C_2H_5Cl)
nitrogen excretion products — urea (CON_2H_4), uric acid ($C_5O_3N_4H_4$)
sugars — glucose ($C_6H_{12}O_6$), etc.

2.4 POLARIZED COVALENT BONDS AND HYDROGEN BONDS

In section 2.2, we introduced the covalent bond using chlorine as an example (I)—

In this case the electronegativity of the two participating atoms is the same and the distribution of electrons between the two atoms will be uniform. However, in a covalent bond between atoms of different electronegativities, the shared electrons tend to be displaced towards the more electronegative atom, giving a polarized covalent bond. For example in the OH bonds of the water molecule (H_2O), where the electronegativities of H and O are 2.1 and 3.5 respectively (see section 1.6 and Table 1.5), the bonding electrons are slightly displaced towards the O atom, giving a partial separation of charge (II, where δ indicates a fractional charge—the angled shape of the water molecule is discussed in section 2.7). Thus we should think of the covalent bond between an oxygen and a hydrogen atom as having some ionic character.

Overall, purely covalent bonds are formed only between atoms of the same, or very similar, electronegativity (e.g. $C-C$, $C-H$ bonds), whereas purely ionic bonds are formed only between highly electropositive metals and highly electronegative non-metals.

Because of the polarization of the $O-H$ bond in water molecules, the partial negative charge on the O atom of each molecule will be attracted to the partial positive charge on the nearest neighbouring water molecule, giving a network of *hydrogen bonds* between the molecules of liquid water—

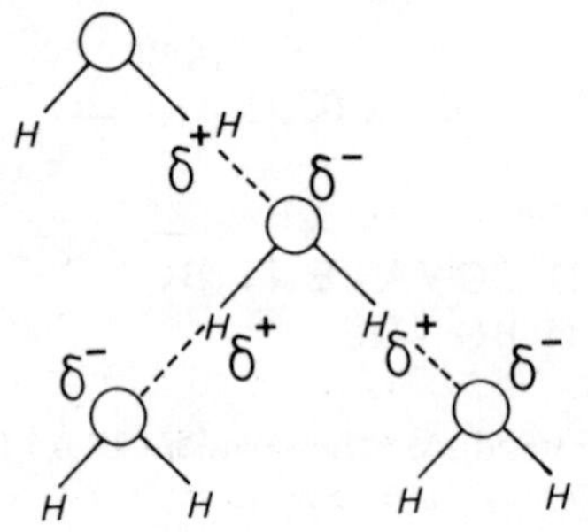

In more general terms, a hydrogen bond forms between a hydrogen atom (carrying a fractional positive charge) on one molecule and a more electronegative atom (not necessarily oxygen, carrying a fractional negative charge) on another molecule.

As we shall see in Chapter 9, the hydrogen bonds of liquid water are of crucial importance for the existence of life on Earth. They are also important in maintaining the shapes and activities of vital biochemical molecules such as enzymes and nucleic acids (Figures 14.1 and 14.3).

2.5 PROPERTIES OF IONIC AND COVALENT COMPOUNDS

Compounds resulting from ionic bonding tend to differ in their properties from covalent compounds. These differences are due mainly to the absence of electrical charges in covalent compounds, although as we saw in section 2.4 the polarization of covalent bonds may modify these differences.

(a) Molecules

When KCl, which is made up to K^+ and Cl^- ions, solidifies from the molten state, each K^+ ion will tend to attract all of the neighbouring Cl^- ions, and vice versa. No distinct molecules, with one K^+ ion tightly bound to one Cl^- ion, are formed; instead, as many K^+ ions as possible (6) cluster round each Cl^- ion, and as many Cl^- ions as possible (6) cluster round each K^+ ion.

In contrast, covalent bonding does result in the formation of molecules, for example, the diatomic molecule of chlorine, Cl_2, in which two atoms are bound tightly together but very loosely to any other chlorine atoms. When covalently bonded substances solidify, the forces holding the molecules together tend to be very weak (van der Waals) forces.

As we shall see in Chapter 4, these weak intermolecular forces (between molecules) and strong intramolecular forces (between the atoms within each molecule) result in solids of low physical strength and low melting point and boiling point. In contrast, ionic solids tend to have much greater strength and higher melting and boiling points. However, there are some notable exceptions to this rule; they include substances like water, with strong intermolecular hydrogen bonds (Chapter 9) and materials like diamond whose network of covalent bonds gives extremely high mechanical strength and melting point (section 4.2).

(b) Solubility

Ionic solids dissolve readily in polar liquids, like water, which are partly charged. For example, when a KCl crystal is placed in water, the fractional negative charges on the water molecules tend to attract K^+ ions and pull them into solution (III) where they exist as hydrated cations (IV)—

III

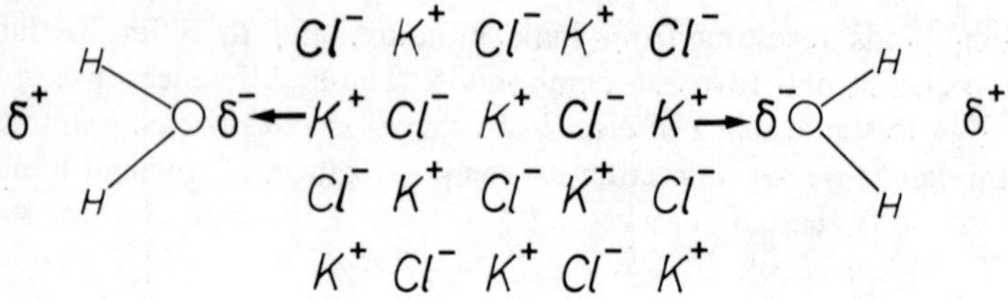

IV

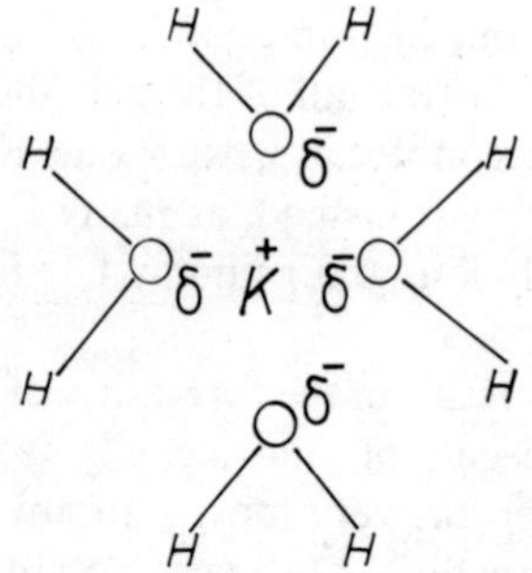

Covalently bonded substances do not contain charged ions and tend to be soluble in non-polar solvents such as alkanes, benzene, etc. (Solubility is treated in much greater detail in Chapters 5 and 9.)

(c) Electrical Conductivity

The presence of charged ions in ionic substances also accounts for the fact that when they are dissolved in water or melted, the liquid will conduct electricity. Normally, electricity passes along metal wires as a flow of electrons; however, the passage of electricity through an ionic solution or melt occurs because the positive ions (cations) move towards the negative pole (cathode) while the negative ions (anions) move towards the positive pole (anode) (Figure 2.1). Thus the current is carried by ions in solution rather than electrons. Because of this property, ionic substances can be described as electrolytes whereas covalent compounds are nonelectrolytes.

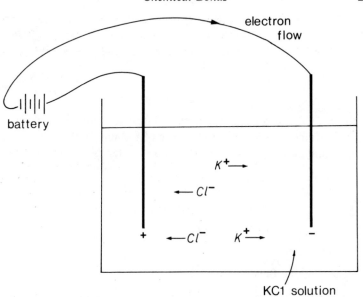

FIGURE 2.1. The flow of electric current through a solution of KCl due to the migration of cations to the negative cathode and anions to the positive anode.

2.6 OTHER BOND TYPES

(a) The Dative Bond

Single covalent bonds are formed when two atoms share a pair of electrons. In the cases we have studied, one electron is contributed by each partner but there are situations where both electrons of a single covalent bond are contributed by one of the participating atoms. This type of bond is called a dative covalent bond, although once formed, it is indistinguishable from other covalent bonds.

The formation of ammonium and hydronium ions in solution gives good examples of the dative bond. Ammonia, NH_3, exists as a molecular gas with the structure as shown (V). The nitrogen atom has a complete outer shell of eight electrons (an octet) and is, there-fore, stable. However, there exist in water a certain number of hydro-gen ions or protons, H^+, lacking electrons. In aqueous solution, an ammonia molecule will form a dative covalent bond with a proton, by contributing a pair of electrons, to give an ammonium ion (VI).

V

$$H \overset{\overset{\circ\circ}{\times}}{\underset{\underset{H}{\circ\times}}{N}} H$$

VI

$$H^{+} + H \overset{\overset{\circ\circ}{\times}}{\underset{\underset{H}{\circ\times}}{N}} H \longrightarrow H \overset{\overset{H}{\circ\circ}}{\underset{\underset{H}{\circ\times}}{N}} H \;^{+}$$

In this way, the nitrogen has maintained its octet while the hydrogen ion has attained the electronic structure of Helium. In the same way, hydronium ions can form, even in pure water (see Chapter 9) —

$$H^{+} + \overset{\overset{\circ\circ}{\times}}{\underset{\underset{H}{\circ\times}}{O}} H \longrightarrow \overset{\overset{H}{\circ\circ}}{\underset{\underset{H}{\circ\times}}{O}} H \;^{+}$$

(b) The Metallic Bond

Most metals have one or two electrons in the outermost shell of their atoms. Since metals are also highly electropositive, these electrons are not tightly bound to the atom. In the solid metal, the atoms pack closely together while their loose outer electrons (1 or 2 per atom) are contributed to a 'shared pool of electrons'. Thus we can consider a sample of solid iron to be a tight framework of iron ions, surrounded by a 'cloud' or 'sea' of electrons. This 'cloud' of shared electrons binds the iron atoms together.

This is obviously an oversimplified theory of metallic bonding but it allows us to understand certain important properties of metals. For example, high electrical conductivity in metals is possible due to the looseness and mobility of the 'cloud' of electrons.

2.7 THE SHAPES OF MOLECULES

As well as having distinct sizes, molecules also have definite 3-dimensional shapes. We can illustrate this by looking at compounds of C, N, O and F with H.

CH_4	Methane
NH_3	Ammonia
H_2O	Water
HF	Hydrogen fluoride

The C, N, O and F atoms in these compounds have achieved a stable octet of outer electrons by bonding with hydrogen. In each molecule, the eight electrons of the octet are grouped into four pairs of electrons. In methane each pair constitutes a single covalent bond (VII), whereas in hydrogen fluoride, three of the pairs are unshared. Clearly these electron pairs, being negatively charged, will repel one another till they are arranged round the atom as far away from one another as possible. This will cause the electron pairs, shared or unshared, to be directed towards the corners of a tetrahedron.

VII

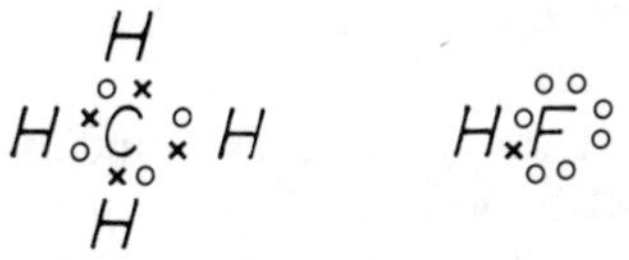

VIII

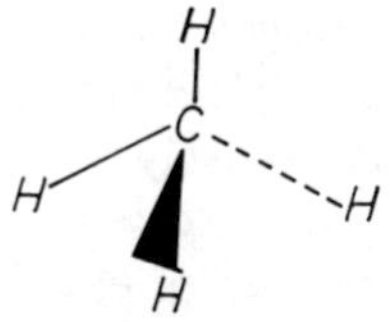

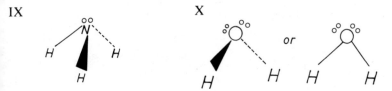

For example, in methane, we have a tetrahedral molecule (VIII) where continuous lines are in the plane of the paper, broken lines indicate bonds directed away from the reader and wedge-shaped lines indicate bonds directed towards the reader. This molecule is regular and symmetric; the angle between any H−C−H is 109·5°.

The electron pairs in ammonia are directed in the same way (IX) but here, one of the electron pairs is not shared and the molecule is, therefore, pyramidal.

Again, in water, the electron pairs are directed towards the corners of a tetrahedron but only two pairs constitute bonds (X), giving a planar molecule with an H−O−H angle of 104·5°.

IX

X

or

Finally, the hydrogen fluoride molecule is linear—

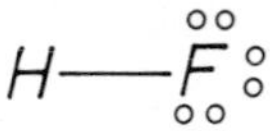

In this way, it is possible to work out the shape of many simple molecules. Molecular shape becomes particularly important in

organic chemistry and biochemistry where compounds are made up
of long chains of tetrahedral carbon atoms.

2.8 VALENCY

Valency is a term used in different ways. For our purposes, we shall
use valency to indicate the number of bonds formed by the atoms
of an element, for example carbon (see section 2.7), whose atoms
form four single bonds with hydrogen atoms, has a valency of 4,
nitrogen has a valency of 3, oxygen 2 and fluorine 1. Where double
or triple bonds are encountered, these are counted as 2 or 3 bonds
respectively.

   In the case of ionic compounds we can use valency to indicate the
number of electrons lost or gained by atoms of an element when
forming the compound. This value will be the same as the size of
charge on the ion of the element. For example, in KCl, each ion will
have a valency of 1, whereas in $CaCl_2$, Ca has a valency of 2 and
Cl, 1.

EXERCISES

Draw 'dot diagrams' to illustrate:
(1) ionic bonding in NaCl, CsF and $MgCl_2$;
(2) covalent bonding in H_2, HCl and Br_2;
(3) bonding in $Ca(OH_2)$.

CHAPTER 3
STATES OF MATTER

Chemical substances can exist in three different states—solid, liquid and gas. In the next three chapters, we shall look at some important features of each state of matter but before we can do that we must establish the differences between the three states of matter.

If we place a stone (solid state) in a glass beaker, it occupies some of the volume of the beaker but does not change in shape. On the other hand, if we pour some liquid water into the beaker, the water changes in shape to fit the volume. Finally, if we start with a vacuum in the beaker and then admit some hydrogen gas, the gas shapes itself to the beaker as does water, but it does not have a definite volume limited by a surface. Instead, the hydrogen gas expands to fill the beaker and the molecules at the top of the beaker mix freely with the air of the room.

These observations allow us to define each state:

(i) *Solids* have definite volumes and shapes, bounded by fixed surfaces;

(ii) *Liquids* have definite volumes but indefinite shapes, bounded by flexible surfaces;

(iii) *Gases* have indefinite volumes and indefinite shapes. Volumes of gases do not have surfaces.

3.1 CHANGES OF STATE

Most substances can exist in each of the solid, liquid and gaseous states. For example, water may be ice, liquid water or water vapour, depending upon the conditions. Normally we are familiar with a substance in one state only (solid iron, liquid mercury, gaseous oxygen) but each substance can exist in other states. If we place solid iron in a sufficiently hot furnace, it will become liquid (melt); if we warm mercury in an open vessel, it will become gaseous (vapourize); if we expose oxygen to extreme cooling, we obtain liquid oxygen by condensation. These events (melting, condensation, etc.), which are called changes of state, occur at constant temperatures if the substance is pure.

The temperature of a body (solid, liquid or gas) is a measure of the velocity of movement of the atoms or molecules of the body, i.e. the kinetic energy of the particles. We can increase the kinetic energy of these particles by supplying energy in the form of heat; the more heat we supply, the faster the particles move and the higher will be the temperature of the body.

In the solid state, the atoms of a piece of iron, for example, are tightly packed together and vibrating very slowly. If we heat the metal, the temperature rises as the atoms begin to vibrate more rapidly. If more heat is supplied, the vibrations become more and more violent until a temperature is reached where the atoms are moving so vigorously that they 'escape' from the solid surface. This causes the solid to lose its definite shape and become a liquid. The temperature of the metal remains steady at the melting point until the applied heat has caused all the solid to melt.

In the liquid state, the atoms are not packed in a tight, orderly way. Because they are free to move about randomly throughout the volume of the liquid, heating causes the atoms to rush about more and more rapidly and randomly within the surface of the liquid. At a second constant temperature, the boiling point, the atoms are moving so quickly that they begin to escape from the liquid surface and spread into the surrounding air. The temperature remains steady throughout the change of state but, thereafter, heating results in a rise in temperature due to the increased velocity of movement of the atoms in the gaseous state.

In passing, we can see that the heating of a solid, liquid or gas results in expansion because the particles in the body require more and more space for vibration and free movement.

It is important to remember that the atoms and molecules in a liquid are not all moving at the same velocity at any instant. For example, the collision of a pair of moving atoms may result in one being temporarily stationary while the other moves off at a velocity equal to the combined original velocities of the two atoms. Such effects can be clearly demonstrated with the balls on a billiard table. Consequently, since collisions between particles occur continuously, we should think of the temperature of a liquid (solid or gas) as a measure of the mean velocity of the particles where as many particles will have a velocity higher than the mean as lower.

This explains why changes of state are not instantaneous; the faster particles escape first, leaving the slower particles to be accelerated to the critical velocity for escape by the heat supplied. It also explains the common observation that water evaporates from soils, lakes and even drying washing, at temperatures far below the boiling point of water (100°C). Here again, the very few molecules which are moving at the 'escape velocity' evaporate first. This evaporation causes a loss of kinetic energy to the gas and, therefore, cools the remaining liquid water. However, as long as the water is warmed back to its original temperature by the air or by absorption of solar radiation, there will be a continuous loss of the most energetic molecules. The time taken for complete evaporation of a volume of water at a temperature below its boiling point, will, therefore, depend upon the air temperature, wind speed, and the supply of solar radiation.

When some solids are heated, their atoms or molecules pass directly into the gas state without passing through the liquid state. This change of state, which is relatively uncommon, is called sublimation. The process can be clearly demonstrated by the sublimation of iodine; on heating, the black crystals of iodine release dense purple fumes of iodine vapour. Similarly, snow can change directly to water vapour under intense solar radiation, for example into the dry air of the Arctic.

The slow loss of molecules from a solid directly to the gas state at low temperatures can be very useful in pest control. For example, the release of molecules from a solid insecticide in a confined space can give protection from insect damage to stored materials such as agricultural products (e.g. grain). Another familiar example is the use of naphthalene (in 'mothballs') to protect stored cloth from the clothes moth.

3.2 TEMPERATURE, TEMPERATURE SCALES AND THERMOMETERS

Temperature is a very important factor in all branches of science. To give a few relevant examples—air and soil temperatures control the growth of plants; the body temperatures of livestock are useful in the diagnosis of disease; the burning temperature of a gas mixture determines its usefulness in cutting and welding metals.

In order to measure and compare the temperatures of soils, animals or flames, we require accurate thermometers calibrated over the appropriate range of temperature. The standard temperature scale for most scientific work is the Centigrade or Celsius scale where the melting point of pure ice and the boiling point of pure water (both constant temperatures) are defined as 0°C and 100°C respectively at 1 atm air pressure. Thermometers can be calibrated for use within this range using these fixed points; for measurements below 0°C or above 100°C, additional changes of state may be used to extend the range of centigrade thermometers.

After the development of the centigrade scale, it was discovered that the lowest possible temperature was −273°C, i.e. 273 degrees below the melting point of ice. This led to the development of an absolute scale of temperature which retained the same size of units as the centigrade scale but which had its zero point at −273°C. The scales are, therefore, related as follows:

```
°A    0     1     2  ----- 272 273 274 ----- 372 373 -----
°C −273 −272 −271 ----- −1  0   1  -----  99 100 -----
                                              etc.
```

Although not convenient for scientific work, the Fahrenheit scale is still generally used in everyday life. For example, to most people the air temperature on a hot day will be between 80 and 100°F rather than 27–38°C. When devising his scale in 1714, Fahrenheit chose as his zero point the coldest temperature he could obtain (by mixing equal quantities of snow and ammonium chloride). His 100°F was the temperature of human blood, now known to be 98·4°F. This resulted in temperatures of 32°F for the melting point of water and 212°F for the boiling point. Since one degree Fahrenheit is equal in magnitude to $\frac{5}{9}$ of a degree centigrade, the three scales are related as follows:

$$°C = °A - 273$$

$$°C = (°F - 32) \times \frac{5}{9}$$

$$°F = \left(°C \times \frac{9}{5}\right) + 32.$$

Since liquids expand on heating, the volume of a fixed mass of liquid can be used as a convenient index of temperature. This is the basis of the familiar glass/mercury thermometer in which variations in temperature can be measured by the expansion and contraction of the liquid metal along a glass tube of uniform internal bore.

Mercury has a unique combination of physical and chemical properties which make it an outstandingly good liquid for general purpose thermometers. As well as ease of visibility and a high thermal conductivity, these properties include:

(i) *High Density* Because of its high density (13·6 g ml^{-1} compared with 1 g ml^{-1} for water) mercury has a small (but regular) increase in volume per unit increase in temperature. Consequently, glass/mercury thermometers can be compact and portable.

(ii) *Liquid Range* Mercury exists as a liquid between its melting point, $-39°C$, and its boiling point, $357°C$. This liquid range conveniently covers the majority of temperatures encountered in scientific work.

(iii) *Surface Properties* The combination of strong forces between the atoms of liquid mercury and very weak attraction between the atoms and glass surfaces (see Chapter 10) ensures that the liquid column does not break as it advances and retreats and does not leave a 'tail' of mercury to obscure temperature readings.

Glass/mercury thermometers cannot be used at extremes of temperature beyond the liquid range of mercury or in field work where the glass is liable to be broken. Under these circumstances it is convenient to use thermocouples or thermistors, whose electrical resistances vary with temperature. When attached to a suitable power source and recorder, these sensors can give continuous records of temperature over prolonged periods.

EXERCISE

Consult reference books to discover whether the following substances would be suitable as substitutes for mercury in thermometers: Ethanol; Carbon tetrachloride; Benzene; Iodine.

FURTHER READING

1. HAY R.K.M. (1976) The temperature of the soil under a barley crop. *J. Soil Sci.*, **27**, 121–8 (illustrates the use of automatic recorders and thermistors in long-term monitoring of environmental temperature).
2. ROSSOTTI H. (1975) *Introducing Chemistry*, Ch. 15. Penguin Books, Middlesex.

CHAPTER 4

THE SOLID STATE

As we saw in the last chapter, the atoms, ions or molecules in solids are tightly packed together. According to the regularity of this packing, solid substances can be classed as crystalline or amorphous.

Crystalline solids (Crystals) have regular three-dimensional shapes with flat faces, straight edges, sharp vertices and planes of weakness parallel to their faces. This regularity of shape is a direct consequence of the regularity of packing of the particles in the solid.

Amorphous solids do not have regular three-dimensional shapes.

Most of the substances studied in chemistry are crystalline although a few, such as proteins and clay-sized minerals, may be amorphous. In this chapter we shall investigate some relationships between the crystal structure (arrangement of particles) of solid substances and their bulk properties.

4.1 CRYSTAL STRUCTURES

Ionic and metallic substances have close-packed crystal structures, as shown for NaCl in Figure 4.1. In NaCl the maximum number of negative chloride ions (6) is packed round each positive sodium ion, and the maximum number of sodium ions (6) is packed around each chloride ion. This arrangement is repeated throughout the crystal giving a uniform structure held together by strong ionic bonds. Many ionic compounds, including KCl, have similar cubical crystal structures.

Because of uniform, strong bonding in their crystals, ionic solids tend to be mechanically strong and to have high melting points since a great deal of energy must be supplied to enable ions to escape from the solid surface. For example, NaCl melts at 801°C and KCl at 776°C.

The structures of ionic solids also account for the characteristically regular shapes of their crystals. In our example, NaCl crystals are cubic simply because the arrangement of Na and Cl ions in the crystals is cubic; the crystals split naturally along planes of weakness

32

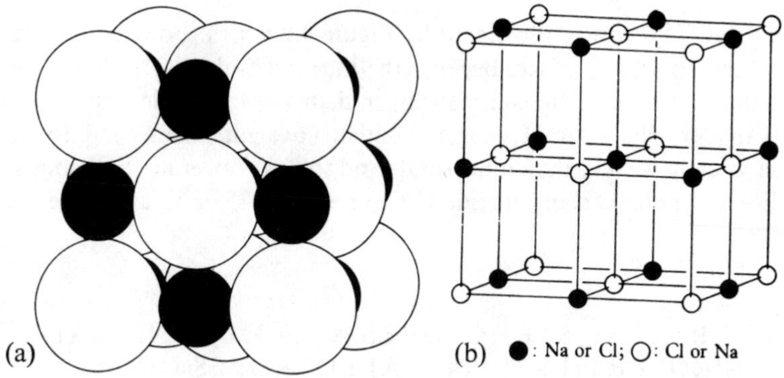

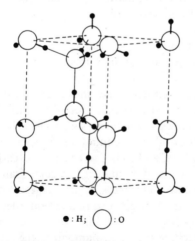

(a) (b) ● : Na or Cl; ○ : Cl or Na

FIGURE 4.1. The crystal structure of sodium chloride, NaCl. (a) Showing the close-packed arrangement, where the sodium atoms are represented by the smaller spheres, and (b) stressing the relative positions of the ions in the structure. (From *An Introduction to Crystal Chemistry*, 2nd Edition, by R.C. Evans (1966). Cambridge University Press.)

between the layers of ions to give regular faces, edges and vertices (Figure 4.1).

Covalently bonded substances are also normally crystalline in the solid state. Their molecules are arranged in a three-dimensional pattern which, as in ionic solids, repeats regularly throughout the

● : H; ○ : O

FIGURE 4.2. The hexagonal crystal structure of ice showing the regular three-dimensional arrangement of water molecules held together by hydrogen bonds. (From *An Introduction to Crystal Chemistry*, 2nd Edition by R.C. Evans (1966). Cambridge University Press.)

crystal. This can be illustrated by Figure 4.2 which shows one of the crystal structures of ice. Because the intramolecular forces (covalent bonds) in their crystals are stronger than the intermolecular forces (van der Waals or Hydrogen bonds), covalent solids tend to be physically weaker than ionic solids and to have lower melting points. For example, carbon dioxide (CO_2) melts at $-56\cdot6°C$ and water at $0°C$.

4.2 CRYSTAL STRUCTURES AND MECHANICAL PROPERTIES. 1. THE ALLOTROPES OF CARBON

One of the best examples of the relationship between crystal structure and bulk mechanical properties is provided by the allotropes of carbon.

Allotropes are different physical forms of the same solid element.

Solid carbon can exist in a number of different forms. Many of these are amorphous and impure products of the burning of carbon-containing fuels without adequate oxygen (e.g. charcoal from wood, coke and soot from coal, lampblack from candles). However, there are two naturally occurring pure and crystalline allotropes of carbon:

(1) *Diamond* Brilliant, colourless and hard octahedral crystals. Formed naturally or artificially at very high pressures and temperatures.

(2) *Graphite* Dull black flaky crystals formed at lower pressures.

These differences in appearance result from different crystal structures:

Diamond (Figure 4.3a) As we saw in section 2.7, the carbon atom can form four single covalent bonds directed towards the corners of a regular tetrahedron. In a diamond crystal, each carbon atom forms a single bond with four other carbon atoms, each of which, in turn, is bonded to three others, and so on, so that all the carbon atoms in a diamond crystal are firmly bound in a regular framework. Unlike most covalently bonded substances, no molecules are formed and the bonding is uniform in all directions; in addition, there are no simple planes of weakness (compare with the NaCl structure, Figure 4.1). Because of this crystal structure, diamond is the hardest naturally occurring substance and has a very high melting point (3500°C). Diamonds can, therefore, be used to cut other hard substances such as glass or rock (used as the 'bit' when drilling oil or water wells).

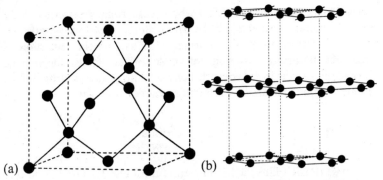

FIGURE 4.3. The crystal structures of the allotropes of carbon. (a) diamond and (b) graphite. (From *An Introduction to Crystal Chemistry*, 2nd Edition, by R.C. Evans (1966). Cambridge University Press.)

Graphite (Figure 4.3b) The carbon atoms in graphite crystals also have a valency of four. However, each atom forms short and strong covalent bonds with three other carbon atoms and one longer and weaker bond with a fourth carbon atom. As Figure 4.3b shows, this results in a series of flat layers of tightly bonded carbon atoms held together in stacks by weaker inter-layer bonds. We can think of each layer as being a very large two-dimensional molecule.

Graphite is a soft substance due to its distinct planes of weakness which allow the layers to move over one another. Because of this property, graphite is used in pencil leads and as a lubricant, especially at high temperatures which cause the decomposition of petroleum oils. When we write with a pencil, we rub off layers of graphite which mark the paper. When graphite is used to lubricate machinery, friction is reduced by the sliding of layers of carbon atoms.

Thus, according to the arrangement of the atoms, carbon crystals may be extremely hard, or soft and flaky (Table 4.1).

4.3 CRYSTAL STRUCTURES AND MECHANICAL PROPERTIES. 2. METALS

Iron and its alloys (the ferrous metals) are by far the most important metals used in engineering and construction. Indeed, iron accounts for more than 90% by weight of all metals used in the world today. Ferrous metals can be classed into three groups:

(a) Pure iron (forged iron, wrought iron, ingot iron);

(b) Iron containing 2·5–5% carbon (cast iron, pig iron);

(c) Iron containing 0·2–1·5% carbon (steels of various types).

Iron and steel production normally begins with the reduction of iron ores (naturally occurring iron oxides or carbonates) in the blast furnace. The ore is mixed with coke and limestone and heated to 1800°C. Iron is released by reduction reactions such as:

$$Fe_2O_3 + 3C \rightarrow 2Fe + 3CO \qquad \text{(see Chapter 7).}$$

$$\underset{\substack{\text{Haematite}\\ \text{ore}}}{} \quad \underset{\text{Coke}}{} \quad \underset{\text{Iron}}{} \quad \underset{\substack{\text{Carbon}\\ \text{monoxide}}}{}$$

It can be drawn off as the liquid metal (m.p. 1528°C) and poured into moulds to give cast iron objects. Many of the impurities in the ore, such as silicate minerals, which would impair the properties of the cast iron, are absorbed by the limestone to give a liquid material of low density called basic slag. As this floats on the surface of the liquid metal, it can be removed when withdrawing the iron. Basic slag is a useful phosphorus fertilizer for acid soils (see section 15.2).

In spite of the refining action of limestone in the blast furnace, cast or pig iron still contains many impurities, including 2·5–5%C as well as silicon, manganese, phosphorus and sulphur compounds. In the manufacture of steel, molten pig iron from a blast furnace is purified by one of a number of different processes (Bessemer, Open hearth, etc.). The exact quantities of carbon and other ingredients are then added to give steel of the required quality. Wrought iron, which is no longer an important material, can be prepared by low temperature purification of pig iron.

The three classes of ferrous metal have distinctly different mechanical properties which are related to differences in the arrangement of their atoms. Before discussing these differences, it is necessary to present some useful definitions:

(1) *An Alloy* is a mixture of two or more metals or a mixture of a metal with a non-metal. In some cases the ingredients may react together to form an intermetallic compound.

(2) *The Hardness* of a solid substance is a measure of its resistance to scratching, wear or penetration. By observing the ability of one solid substance to scratch or cut the surface of other substances, it has been possible to draw up scales of hardness (e.g. Table 4.1). In materials testing, the hardness of a solid substance may be assessed more exactly by measuring the size of impression formed when a diamond point is pressed into the surface of the solid.

TABLE 4.1. The Moh scale of hardness of minerals.

Hardness index	Typical mineral
1	Talc
2	Gypsum, Rock Salt
3	Calcite
4	Fluorite
5	Apatite
6	Orthoclase Feldspar
7	Quartz
8	Topaz
9	Corundum
10	Diamond

Thus Graphite with a hardness between 1 and 2 is harder than talc but softer than rock salt.

(3) *The Strength* of a metal is a measure of its resistance to deformation when acted upon by a force. In particular, the *Tensile Strength* of a metal is the maximum pulling load that a sample of the metal can endure without cracking or breaking.

(4) The *Ductility* and *Malleability* of a metal, which are measures of the ease of drawing the metal into a wire or hammering it into a flat sheet, depend upon the strength of the metal. A metal with a high strength has low ductility and malleability (i.e. it resists deformation).

The mechanical properties of ferrous metals are compared qualitatively in Table 4.2; some of these differences can be explained using simple models of the structure of metals.

Most metals, including iron, are so reactive that they exist naturally as compounds and do not normally occur as the 'native' metal (exceptions are gold, silver and copper). Since the familiar metal objects we use have been shaped artificially, metals do not, at

TABLE 4.2. Mechanical properties of ferrous metals.

	Hardness	Tensile Strength	Ductility	Malleability
Pure Iron	Low	Low	High	High
Cast Iron	High	Low	Low	Low
Steel	High	High	Low	Low

first sight, appear to be crystalline according to the definition given on page 32, i.e. they do not have regular shapes with flat faces, straight edges and sharp vertices. However, metals are considered to be crystalline because it can be shown that in the solid state the atoms are arranged in regular three-dimensional patterns or lattices. For example, a bar of pure iron contains a large number of small iron crystals (of variable size, but typically smaller than 0.1 mm) each made up of close-packed layers of iron atoms. If the bar is subjected to a hardness or tensile strength test, the layers of identical iron atoms will tend to slide over one another under the influence of the applied force, leading to permanent deformation of the bar (Figure 4.4). However, the resulting deformation will not be as great as it would be if the bar were a single crystal, because movement will be resisted by those crystals whose layers are not aligned along the direction of sliding.

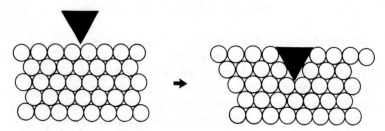

FIGURE 4.4. Schematic diagram of a metal hardness test in which a diamond point, under a standard load, causes a permanent deformation of the metal surface due to the sliding of layers of identical metal atoms.

Because of its tendency to deform easily under mechanical stress, unaltered pure iron is not a useful metal in engineering where strength and hardness are important. To make the metal more useful, it is necessary to insert some 'grit' or irregularity into the structure to reduce the sliding of the layers of iron atoms.

Steels (alloys of Fe and C) are made by the addition of exact small quantities of carbon (normally $0 \cdot 2$–$1 \cdot 5 \%$) to purified molten iron. In the liquid state and in the solid at high temperatures, the added carbon is in true solution (see Chapter 5) and is distributed uniformly throughout the metal. Below 800°C the carbon comes out of solution and becomes concentrated in layers of iron carbide (Fe_3C) which are laid down between crystals of pure iron. By interleaving between layers of iron atoms, iron carbide, which is itself a very hard substance, provides the required 'grit' to increase the 'friction' between iron atoms and reduce slippage. Consequently, steel is harder and stronger than pure iron and more useful as a construction material.

In contrast, there is too much 'grit' in cast iron (up to 5%) where the presence of large fragments of graphite or iron carbide breaks up the regularity of the metal structure. Although cast iron is usually hard, it tends to be brittle, failing at the irregularities in the structure when loaded or under tension. As a result, cast iron is much less useful than steel.

In much the same way, the incorporation of a small proportion (up to 12%) of (larger) tin atoms into a sample of copper can change the metal from a soft, ductile and malleable metal used only for decorative purposes or in plumbing, to bronze, a very hard alloy, formerly widely used in making tools and weapons. The discovery of bronze, probably by the accidental smelting together of copper and tin ores, was one of Man's earliest technological advances.

4.4 SILICATE MINERALS IN ROCKS AND SOILS

Since the elements oxygen and silicon are by far the most common elements in the earth's crust (47% and 28% respectively, by weight), it is not surprising to find that the majority of the minerals occurring in rocks and soils are compounds of Si and O—the silicates.

Like carbon, Si is in group 4 of the periodic table; each of its atoms has four electrons in the outer shell and requires four more electrons to complete an inert gas structure (Ar). On the other hand, oxygen atoms (group 6) have six electrons in their outer shells and require a further two electrons to give the electronic structure of Ne.

The basic unit of all silicates is the Si—O tetrahedron in which a silicon atom is bound by single covalent bonds to four oxygen atoms arranged at the corners of a tetrahedron—

In this way, each Si atom obtains a share in a stable octet of electrons but each oxygen is still one electron short of the neon electronic structure. This deficit in electrons may be made up in a number of ways, leading to a variety of naturally occurring silicate compounds of varying physical and chemical properties.

(a) The Orthosilicates (Independent Tetrahedra)

In the orthosilicate group of minerals, the four electrons required to satisfy each Si—O tetrahedron are supplied by metal atoms, for example, in magnesium olivine, Mg_2SiO_4—

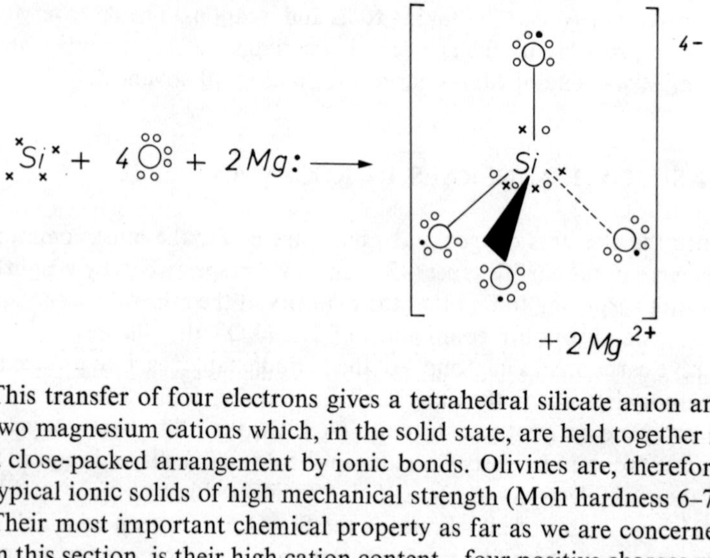

This transfer of four electrons gives a tetrahedral silicate anion and two magnesium cations which, in the solid state, are held together in a close-packed arrangement by ionic bonds. Olivines are, therefore, typical ionic solids of high mechanical strength (Moh hardness 6–7). Their most important chemical property as far as we are concerned in this section, is their high cation content—four positive charges per Si atom.

(b) The Single-Chain Inosilicates or Pyroxenes (single chains of tetrahedra)

In this group of minerals, each silicon atom shares a pair of oxygen atoms with two other silicon atoms, leading to long chains of silicate tetrahedra bound together by Si—O—Si single covalent bonds (i.e. O atoms are shared between tetrahedra). By this condensation of tetrahedra, half of the oxygen atoms achieve a stable octet of electrons and only two electrons per tetrahedron are required from metal atoms. Thus, the magnesium pyroxene Enstatite, $MgSiO_3$, is made up of long chains of negatively charged tetrahedra held together by positive magnesium ions (Figure 4.5).

Because of the very large size of the silicate 'anions' in pyroxenes the crystal structure is less regular than in orthosilicates, with the result that pyroxene minerals are not as strong and hard (Moh hardness 5–6). Pyroxenes have a cation content equal to two positive charges per Si atom.

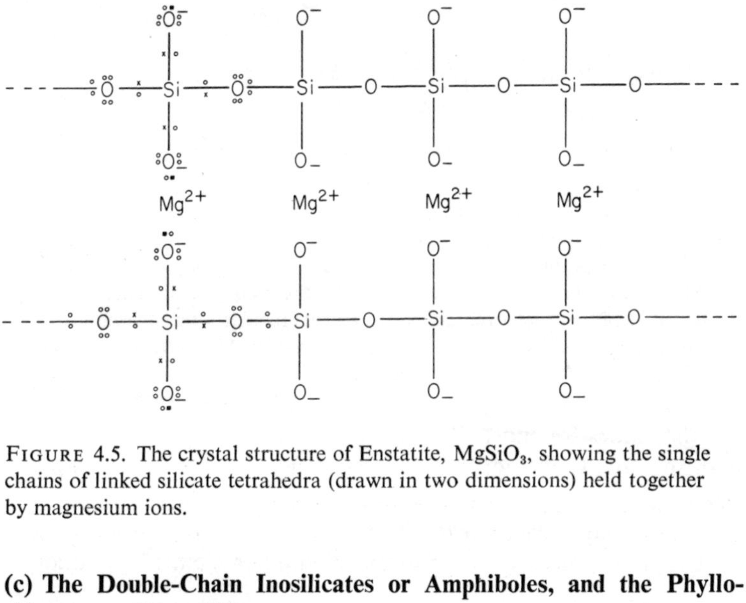

FIGURE 4.5. The crystal structure of Enstatite, $MgSiO_3$, showing the single chains of linked silicate tetrahedra (drawn in two dimensions) held together by magnesium ions.

(c) The Double-Chain Inosilicates or Amphiboles, and the Phyllosilicates or Sheet Silicates.

These minerals result from further condensation of silicate tetrahedra. Amphiboles, which are made up of pairs of parallel pyroxene-like chains connected by $Si-O-Si$ bridges, resemble pyroxenes in their physical properties (Moh hardness 5–6). They normally have complex chemical compositions, e.g. Riebeckite Na_2Fe_5 $(Si_4O_{11})_2$ $(OH)_2$, but their cation content is lower than pyroxenes (equivalent to 1.75 positive charges per Si atom).

In the sheet silicates (Figure 4.6), three oxygen atoms in each tetrahedron are shared with other tetrahedra, resulting in continuous silicate sheets held together by cations. This structure should give a cation content equivalent to one positive charge per Si atom (i.e. one electron required for the fourth oxygen atom which is not shared)

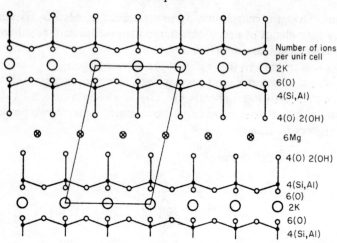

FIGURE 4.6. A cross-sectional representation of an idealized phyllosilicate crystal structure, in which the sheets of silicate tetrahedra (indicated in two dimensions by silicate *chains*) are held together by alternating layers of K^+ and Mg^{2+} ions. The *unit cell*, indicated by the enclosed area, is defined as the smallest part of a crystal structure which repeats regularly throughout the structure. (From *An Introduction to the Rock Forming Minerals*, by W.A. Deer, R.A. Howie & J. Zussman (1966) Longman.)

but the values for most sheet silicates (e.g. micas) are much higher than this due to isomorphous replacement of some Si atoms by Al atoms.

Isomorphous Replacement is the replacement of an atom (or ion) in a crystal structure by an atom of another element (or another

TABLE 4.3. The radii of some metal ions (values in Ångström units, where $1Å = 10^{-10}m$) in relation to their position in the Periodic Table.

Li^+	Be^{2+}				
0·60	0·31				
Na^+	Mg^{2+}			Al^{3+}	$*Si^{4+}$
0·95	0·65			0·50	0·41
K^+	Ca^{2+}	Fe^{2+}, Fe^{3+}	Cu^+		
1·33	0·99	0·80　0·64	0·96		
Rb^+	Sr^{2+}				Sn^{4+}
1·48	1·13				0·71
	Ba^{2+}				
	1·35				

* Si is not strictly a metal (see section 1.6); the value given is for Si in silicate crystals.

type of ion) without changing the crystal structure. This replacement can occur only if the atoms (ions) are of similar size (e.g. Al for Si, Mg for Al, Table 4.3).

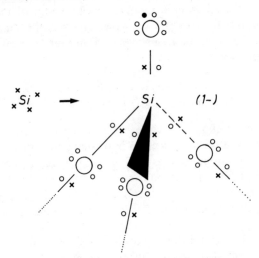

a *Normal tetrahedron in sheet silicate*

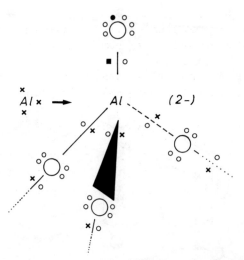

b *Tetrahedron with isomorphous replacement*

FIGURE 4.7. Silicate tetrahedra from phyllosilicates (a) without isomorphous replacement, requiring one extra electron (●) from a cation, and (b) with replacement of Si by Al, requiring two extra electrons (●, ■) from cations, to complete the octet of each oxygen atom.

Thus the Si atom at the centre of a silicate tetrahedron can be replaced by an Al atom whose radius is only a little greater than that of Si (Table 4.3). The space within the tetrahedron is large enough to accommodate the Al atom without great distortion of the crystal structure, but not K or Ca, for example. This substitution of Si by Al occurs to the extent of about one atom in four in micas, as indicated by their formulae:

Muscovite (white mica) $KAl_2(AlSi_3)O_{10}(OH,F)_2$
Biotite (black mica) $K(Mg,Fe)_3(AlSi_3)O_{10}(OH,F)_2$

Although Al is about the same size as Si, it is in Group 3 and, therefore, contains one less electron in its outer shell than Si. Therefore, when an Al atom replaces an Si atom in a silicate tetrahedron, it must obtain a further electron from a cation (Figure 4.7). Thus the cation content of micas is between 1 and 2 positive charges per tetrahedron, depending upon the degree of isomorphous replacement.

Because of the slight distortion of their structures by isomorphous replacement and also because of their well-developed planes of

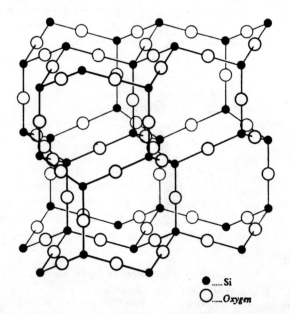

●......Si
○......*Oxygen*

FIGURE 4.8. One of the crystal structures (β-tridymite) of SiO_2. (From *Atomic Structure of Minerals*, by W.L. Bragg (1937). Cornell University Press.)

weakness, sheet silicates tend to be flaky and physically weak (Moh hardness 2.5).

(d) The Tektosilicates or Framework Silicates

When all four of the oxygen atoms in silicate tetrahedra are shared with adjacent tetrahedra, then a crystal structure is formed which is covalently bonded throughout and strongly resembles the structure of diamond (Figures 4.3a and 4.8). Thus quartz (SiO_2) is mechanically strong (Moh hardness 7) and contains no cations.

However, isomorphous replacement can also occur in framework silicates leading to a reduction in strength by crystal distortion and an increase in cation content. For example, orthoclase feldspar $K(AlSi_3)O_8$, which has the same structure as quartz but with one Si atom in four replaced by Al, has a Moh hardness of 6 and a cation content equivalent to 0.25 positive charges per tetrahedron.

TABLE 4.4. The approximate proportions of mineral species exposed to weathering at the earth's surface. (From *Fluvial Processes in Geomorphology* by Luna B. Leopold, M. Gordon Wolman & John P. Miller (1964). W.H. Freeman and Co. © 1964.

Feldspars	30%
Quartz	28%
Micas and Clay Minerals	18%
Limestones ($CaCO_3$)	9%
Iron oxides	4%
Inosilicates	1%
Others (including orthosilicates)	10%

4.5 THE WEATHERING OF SILICATE MINERALS

Soils are formed by the physical, chemical and biological weathering of rocks exposed at the earth's surface, whose mineral content is given in Table 4.4. In general, the largest soil particles (sand and silt > 0.002 mm diameter) are fragments of rock minerals, broken down by physical weathering (action of ice, heat, water, wind, etc.) but relatively unaffected by chemical weathering. Since resistance to physical weathering depends primarily on physical strength and hardness, we should expect from Table 4.5 that the sand and silt fractions of soils would be composed primarily of the hardest silicates —orthosilicates, feldspars and quartz.

TABLE 4.5. Some properties of silicate minerals.

Mineral type	Moh hardness	Cation content (no. of positive charges/tetrahedron)
Olivines (orthosilicates)	6–7	4
Pyroxenes ⎱		2
⎰ (inosilicates)	5–6	
Amphiboles ⎰		1·75
Micas (phyllosilicates)	2·5	1–2
Feldspars ⎱	6	0·25
⎰ (tektosilicates)		
Quartz ⎰	7	0

In contrast, the clay particles (< 0.002 mm) in soils are predominantly new minerals synthesized during the chemical weathering of rock, sand and silt minerals. The first step in this process is the replacement of cations in silicate structures by hydronium ions (H_3O^+, see section 2.6 and Chapter 8) causing distortion of the crystal structure and the loss of silicate tetrahedra into solution. Thus, the cations in silicate minerals are the primary sites of chemical attack; the higher the cation content of a mineral, the more rapidly it will be weathered. Consequently, of the three hardest silicate minerals in rocks, quartz is very resistant and orthosilicates are very susceptible to chemical weathering, with feldspars intermediate.

As a result of this resistance to both physical and chemical weathering (which depends upon its crystal structure) quartz is by far the dominant mineral (up to 90%) in the sand fraction of most soils, although it accounts for a much smaller fraction of the parent rocks. Due to their susceptibility to chemical weathering, the other rock minerals tend to disappear from the sand fraction, leading to the synthesis of silicate clays. For example, an American granite which originally contained 53% feldspar and 32% quartz weathered to give a soil whose sand fraction contained 50% quartz and 28% feldspar. As weathering proceeds, the proportion of quartz will continue to increase.

EXERCISES

(1) Using a 'ball and stick' molecule construction kit, build models of diamond and graphite. Examine their structures from various angles to establish natural planes of weakness.

(2) In section 4.3, we concentrated upon the influence of carbon on the properties of iron and steel. The mechanical properties of iron/carbon alloys also depend upon the other ingredients and upon heat treatments. Draw up a list of ferrous metals giving details of carbon content, other ingredients, uses and mechanical properties.

FURTHER READING

More details of the great variety of possible ionic and covalent crystals structures can be found in:
1. ADDISON W.E. (1963) *Structural Principles in Inorganic Compounds*. Longman, London.
2. EVANS R.C. (1964) *An Introduction to Crystal Chemistry*, 2nd Edition. Cambridge University Press, Cambridge.
The production, structures, properties and uses of metals are fully discussed in:
3. ALEXANDER W. & STREET A. (1976) *Metals in the Service of Man*, 6th Edition. Penguin Books, Middlesex.
For a more advanced discussion of the structure and weathering of silicate minerals, consult:
4. PATON T.R. (1978) *The Formation of Soil Material*. Allen and Unwin, London.
Silicate clays are also silicate minerals, resembling sheet silicates, but interleaved with sheets of aluminium hydroxide. Their structures can be studied in:
5. WHITE R.E. (1979) *Introduction to the Principles and Practice of Soil Science*. Blackwell Scientific Publications, Oxford.
6. GRIM R.E. (1968) *Clay Mineralogy*, 2nd Edition, Ch. 4. McGraw Hill, London and New York (more advanced treatment).

CHAPTER 5
THE LIQUID STATE

Few substances occur as pure liquids in everyday life although mercury in thermometers is a familiar exception. Instead, most liquids are solutions in which one or more substances (solutes) are dissolved in a pure liquid (solvent).

A *solution* is formed when a solute is separated into its constituent particles (molecules, ions, atoms) and dispersed uniformly throughout a solvent.

In this chapter, we shall confine our discussion to aqueous solutions in which the solvent is liquid water. However, both solute and solvent can exist in each of the three states of matter. For example, many alloys consist of one solid metal dissolved in another; similarly, air may be considered to be a solution of oxygen, carbon dioxide and the inert gases in nitrogen as solvent. An understanding of the properties of aqueous solutions is an essential part of all studies of environmental chemistry, because of the central role that aqueous solutions play in environmental processes. For example:

(a) A large proportion of the Earth's total primary production originates from aqueous solution (marine and freshwater ecosystems).

(b) Most biochemical reactions take place in the aqueous environment of plant cells, animal cells and animal body fluids.

(c) Pesticides and some nutrients are applied to crops and livestock as dilute aqueous solutions (e.g. fungicide and herbicide sprays in crop protection; foliar nutrient sprays; treatment of livestock against ticks by spraying or dipping). These agrochemicals can be applied more safely, accurately and uniformly in solution than as dry solids.

(d) Other aqueous solutions include the soil solution, which is normally the only source of inorganic nutrients for plant uptake, and irrigation water, whose solute content must be low to prevent the development of salinity in soils.

In this chapter, we shall consider some of the basic properties of aqueous solutions.

5.1 SOLUTION CONCENTRATION EXPRESSIONS

In the preparation and use of solutions, it is necessary to express the concentration of dissolved solute in an exact way. The simplest form of expression, mass of solute per unit volume of solvent (e.g. grams/litre) is familiar in everyday life. However, in experimental work, where the effects of two or more different solutes are being compared, it is more convenient to use molarity to express solute concentration. Before we can use this expression, it is necessary to define some new terms.

The *Molecular Weight* (*MW*) of a covalently bonded compound is the sum of the atomic masses of the atoms in one molecule of the compound. Thus for urea, CON_2H_4, which has 1 carbon, 1 oxygen, 2 nitrogen and 4 hydrogen atoms in each molecule,

$$MW = 1(12) + 1(16) + 2(14) + 4(1) = 60.$$

Although there are no molecules in ionic compounds the term molecular weight is still used. The MW of an ionic compound is the sum of the atomic masses of the atoms in the *formula* (which expresses the simplest ratio of ions in the compound, i.e. KCl, not K_2Cl_2 or K_3Cl_3, etc.). Therefore, for potassium chloride, KCl,

$$MW = 1(39) + 1(35.5) = 74.5 \text{ (occasionally this is called the formula weight).}$$

Note that the strictly correct term Molecular Mass is never used.

The *Gram Molecular Weight* (*GMW*) of a compound is its molecular weight expressed in grams (rather than daltons) and is the mass of 1 *Mole* of the compound. The GMW of urea and potassium chloride are, therefore, 60 g and 74.5 g respectively.

A *1 Molar* solution of a substance contains 1 gram molecular weight (or 1 mole) of the substance dissolved in 1 litre of solution. For example a 1 M solution of urea contains 60 g urea l^{-1} and a 2 M solution of KCl contains 149 g KCl l^{-1}.

The molarity expression of concentration is particularly useful because it stresses the *number* of molecules (or ions) of the solute per litre. For example, since the mass of each molecule of urea is 60 daltons (approximately 10^{-22} g) and the mass of 1 mole is 60 g, then the number of molecules in a mole is:

$$\frac{60}{10^{-22}} = 6 \times 10^{23}$$

(precise calculation gives 6.02252×10^{23}).

A similar calculation for glucose ($C_6H_{12}O_6$) gives an identical result:

$$\frac{180}{3 \times 10^{-22}} = 6 \times 10^{23}.$$

When this calculation is repeated for other molecular solutes, it always gives the same result (6.02252×10^{23}) which is called the *Avogadro Number*. We can, therefore, conclude that all 1 molar solutions contain the same number of molecules per litre, whatever the solute.

These calculations can also be extended to ionic solutes. For example, 1 mole of KCl contains 1 Avogadro Number of K^+ ions *and* 1 Avogadro Number of Cl^- ions. The total number of particles per mole will, therefore, be 12×10^{23}. In 1 mole of potassium sulphate, K_2SO_4, there are 2 Avogadro numbers of K^+ ions and 1 Avogadro number of SO_4^{2-} ions giving a total of 18×10^{23} ions.

The value of the use of moles rather than masses can be illustrated by a simple experiment in plant nutrition. For example, if we have pure samples of the following nitrogen fertilizer substances:

Anhydrous ammonia	NH_3
Sodium nitrate	$NaNO_3$
Urea	$CON_2H_4,$

how can we find out which of these forms of nitrogen is most easily absorbed by the roots of a crop growing in soil? As a first step we might establish the crop in pots, apply equal amounts of the fertilizer compounds (say 10 g) to different pots and then assess the availability of the nitrogen applied by measuring the amount of nitrogen in the leaves of the crop after a given period of growth. However, in applying the same mass of each fertilizer, we are applying different amounts of nitrogen:

Ammonia GMW $= 1(14) + 3(1) = 17$ g containing 14 g N
Therefore 10 g contains *8.2 g nitrogen*;
Sodium Nitrate GMW $= 1(23) + 1(14) + 3(16) = 85$ g containing 14 g N
therefore 10 g contains *1.6 g nitrogen*;
Urea GMW $= 1(12) + 1(16) + 2(14) + 4(1) = 60$ g containing 28 g N
therefore 10 g contains *4.7 g nitrogen*.

Consequently, we can obtain no useful information from this experiment because the amounts of nitrogen applied, as well as its availability, vary amongst the experimental treatments. If, instead, we apply 1 mole of ammonia (17 g), 1 mole of sodium nitrate (85 g) or 0.5 mole of urea (30 g) to the pots, then the soil receives the same amount of nitrogen (14 g) in each case and differences in leaf N content will reflect differences in the availability of the nutrient.

Other solution concentration terms used widely in chemistry are *Normality* (number of gram equivalent weights per litre) which is considered in Chapter 8, and *ppm* (parts per million, i.e. number of grams per million grams of solution) which is used for very low concentrations of solute, for example for the trace element contents of plants and soils.

5.2 SOLUBILITY

There is a limit to the amount of solute which will dissolve in a solvent and when this limit is reached, the solution is said to be saturated with the solute. The concentration of solute in a saturated solution is the *solubility* of the solute in the solvent, normally expressed as g l^{-1} or g 100 ml^{-1}. The solubility of solutes in water varies widely between zero and over a thousand g l^{-1} according to the chemical nature of the solute, as shown for a selection of compounds in Table 5.1. Clearly, the very low solubilities of some pesticides mean that they cannot be sprayed as aqueous solutions; in particular,

TABLE 5.1. The water solubility of a selection of substances of importance in agriculture and pollution (most values measured at 0°C or 20°C; compiled from a number of sources).

Solute	Solubility (g l^{-1})
Ammonium nitrate	1183
Potassium chloride	347
Calcium sulphate	2
Copper sulphate	316
Sulphur	0 (totally insoluble)
Urea	1000
DDT (insecticide)	0 (totally insoluble)
Dieldrin (insecticide)	$1 \cdot 86 \times 10^{-4}$
Simazine (herbicide)	5×10^{-3}

sulphur, DDT and Dieldrin are normally applied as dry powders. On the other hand, the very high solubility of ammonium nitrate can lead to the pollution of drainage water and watercourses by fertilizer nitrate.

The solubility of solutes is governed by three general rules:

(1) The solubility of most compounds in water increases with increasing temperature;

(2) The solubility of most gases in water decreases with increasing temperature;

(3) Substances tend to dissolve in solvents which are chemically similar to them. Thus ionic substances, which are made up of charged ions, dissolve readily in polar solvents, like water, whose molecules carry partial charges due to the polarization of covalent bonds. Such solutes, which are called *hydrophilic* ('water loving') are relatively insoluble in non-polar solvents. In contrast, un-polarized molecular solutes (including many organic compounds) have very low solubilities in water but dissolve freely in non-polar solvents like benzene, paraffin and petrol. Both solutes and solvents are therefore *hydrophobic* ('water hating').

We shall return to this subject of hydrophobic and hydrophilic substances in Chapter 9 and throughout Part II. However, at this point it should be noted that rule 3 largely explains the great differences in solubility amongst the substances in Table 5.1.

Although most ionic compounds (salts) do dissolve readily in water, there is a group of important inorganic salts which are relatively insoluble, as shown by the following classification:

Soluble Salts (*solubility* > 10 g l^{-1}) —All nitrates and acetates; most chlorides, bromides, iodides and sulphates; all salts of Group 1 metals (Na, K, etc.) and ammonium.

Insoluble Salts (*solubility* < 1 g l^{-1}) —Most hydroxides, carbonates, phosphates and sulphides (except those of the Group 1 metals and ammonium).

Thus, of the three macronutrients in plant nutrition, N, as nitrate or ammonium ions, and K, as potassium ions, are highly soluble, whereas P, as phosphate, is relatively insoluble.

5.3 THE SOLUBILITY PRODUCT

Because hydroxides, carbonates and phosphates are involved in many aspects of soil and water chemistry, it is necessary to have a more precise way of expressing and predicting the water solubility of these sparingly soluble salts. This may be achieved using solubility products.

The *Solubility Product Law* states that in all saturated solutions of a given sparingly soluble salt, the product of the concentrations (moles/litre) of the component ions is the same. This constant product is called the solubility product, and values for some selected salts are shown in Table 5.2. For example, in all saturated solutions of silver chloride, AgCl,

$$\text{Solubility Product } (K_{AgCl}) = [Ag^+] \times [Cl^-] = 1.6 \times 10^{-10},$$

where *square brackets indicate molar concentrations*. Similarly, for all saturated solutions of aluminium hydroxide, Al $(OH)_3$,

$$K_{Al(OH)3} = [Al^{3+}] \times [OH^-] \times [OH^-] \times [OH^-]$$
$$= [Al^{3+}] [OH^-]^3 = 1 \times 10^{-33}.$$

Note that each ion in the formula is involved in the product.

The lower the value of K, the lower is the solubility of the salt. Thus hydroxyapatite, which is a major phosphate mineral in rocks, is effectively insoluble in water (Table 5.2).

Although the solubility product law refers to solutions of one solute only, under certain conditions we can use the law to predict the results of adding a second solute. For example, if the second salt

TABLE 5.2. The solubility products of some sparingly soluble salts, mainly at 18–25°C (compiled from a number of sources).

$CaSO_4 \cdot 2H_2O$	$2 \cdot 4 \times 10^{-5}$
$Ca(OH)_2$	8×10^{-6}
$CaHPO_4$	$1 \cdot 3 \times 10^{-8}$
$CaCO_3$	$4 \cdot 8 \times 10^{-9}$
AgCl	$1 \cdot 6 \times 10^{-10}$
$Fe(OH)_2$	1×10^{-15}
$Al(OH)_2(H_2PO_4)$*	$2 \cdot 8 \times 10^{-29}$
$Al(OH)_3$	1×10^{-33}
$Fe(OH)_3$	1×10^{-38}
$Ca_{10}(PO_4)_6(OH)_2$†	$1 \cdot 5 \times 10^{-112}$

* Variscite—an important phosphate mineral in soils.
† Hydroxyapatite—one of the minerals in rock phosphate.

does not contain an ion common to the first (e.g. $CaCO_3$ and $AgCl$) then each salt will tend to dissolve as if the other were not present.

In contrast, if the two salts do have a common ion, then the addition of the second salt may influence the solubility of the first. As a simple example, if we add some solid KOH to a saturated solution of $Al(OH)_3$ then the hydroxide ion concentration in the solution $[OH^-]$, rises due to the high solubility of the KOH. However, according to the solubility product law, $[Al^{3+}] [OH^-]^3$ must remain constant and therefore $[Al^{3+}]$ must decrease, i.e. Al^{3+} ions come out of solution, giving a precipitate of solid $Al(OH)_3$. Thus the addition of KOH causes a lowering of the solubility of aluminium ions.

This phenomenon occurs in soils where Al^{3+} ions, which are highly toxic to most plant species, become less soluble as the hydroxide ion concentration of the soil solution rises. Thus, only those species whose roots are tolerant of Al^{3+} ions can grow successfully in soils with low hydroxide ion concentration (pH < 5, see chapter 8). On the other hand, crop plants, which are generally sensitive to Al^{3+} ions, are grown in agricultural soils which are maintained at moderate hydroxide ion concentrations (pH 5.5–7.0) by liming.

Overall, it is important to stress that solubility products can be used in environmental chemistry only with great care. Firstly, since most naturally occurring aqueous solutions contain more than one sparingly soluble salt, common ion effects must be taken into account. Secondly, the presence in solution of certain organic compounds (chelating agents) can result in large increases in the solubility of ionic solutes. For example, the molecules of fulvic acid, a soluble component of soil humus, can 'wrap around' lead ions, thereby rendering them soluble and subject to leaching. This is an important process in the mobilization of heavy metal pollutants. In the same way, the eluviated (bleached) E horizons of podzols are caused by enhanced solubility and leaching (cheluviation) of Fe^{3+} and Al^{3+} ions due to polyphenol chelating agents released by heath and coniferous vegetation. The synthetic chelating agent EDTA (whose structure is given in section 14.1) is used in the chemical analysis of sparingly soluble metal ions.

5.4 DIFFUSION AND OSMOSIS

In an aqueous solution, both the water molecules and the atoms, ions or molecules of the solute are in constant, rapid and random motion.

Consequently, any differences in solute concentration which might occur within the solution are quickly abolished by the random movement of solute particles. This movement is called diffusion.

The process of diffusion can be illustrated as follows. The volume of a beaker is divided into two equal parts (A and B) by a watertight partition, A containing 50 ml of 1 M NaCl solution and B containing 50 ml of 2 M NaCl solution. Because there are twice as many Na^+ and Cl^- ions in B than A, the random movement of the solute results in twice as many Na^+ and Cl^- ions per second colliding with the B side of the partition than the A. Consequently, when the partition is removed, twice as many ions per second pass from B to A than pass from A to B and the difference in concentration decreases rapidly to give 100 ml of 1.5 M NaCl solution. Na^+ and Cl^- ions have diffused from a region of high concentration (B) to a region of low concentration (A) by random movement. This movement can be described as diffusion down a concentration gradient.

The diffusion of solute particles from a region of high solute concentration (c_2) to a region of low concentration (c_1) can be described by a form of Fick's First Law:

$$\text{Diffusion Rate} = AD\,\frac{c_2 - c_1}{l}$$

where A is the cross-sectional area through which diffusion is occurring, l is the distance between c_1 and c_2, and D (the diffusion coefficient of the solute in water) is the rate of diffusion when

$$\frac{A(c_2 - c_1)}{l} = 1$$

(e.g. a concentration gradient of 1 mole l^{-1} cm^{-1} and a cross-sectional area of 1 cm × 1 cm.)

D values can, therefore, be used as a measure of the relative rates at which different solutes diffuse down a concentration gradient. For example, Table 5.3 gives D values for nitrate, phosphate and potassium ions in solution and in moist soil. The values are much lower, and, therefore, diffusion is much slower in soil because of the long and 'tortuous' diffusion pathway though the water films round soil particles (since ionic diffusion does not occur through air filled soil pores or through the solid soil particles). The value for phosphate is especially low because phosphate ions tend to precipitate in the soil as they pass along the diffusion path.

TABLE 5.3. Diffusion coefficients for ions in solution and in moist soil at 20–25°C (cm² s⁻¹). (Compiled from a number of sources by R. Scott Russell in *Plant Root Systems*. © 1977. McGraw-Hill Book Co. (UK) Ltd.)

	Solution ($\times 10^{-5}$)	Soil ($\times 10^{-5}$)
Nitrate	1·92	0·5
Potassium	1·98	0·01–0·24
Phosphate	0·89	0·0005–0·001

(Note that the diffusion coefficient of an ion in solution is related to its molecular weight; for example, the phosphate ion is both the heaviest and the slowest of the three ions.)

Because of lower densities and higher particle velocities, diffusion rates are much higher in the gas state than in aqueous solution (as shown in Table 6.2). In contrast, diffusion is extremely slow in solids.

Returning to our beaker, if we replace the partition between A and B with a membrane which is permeable to Na^+ and Cl^- ions, then the equalization of NaCl concentration in A and B occurs by diffusion through the membrane. However if we replace the permeable membrane with a *semi-permeable* membrane, which is permeable to water but not to solutes, then the Na^+ and Cl^- ions cannot diffuse through the membrane from B to A to give equalization of solute concentration. Instead, water, which is at a higher 'concentration' in A, diffuses through the membrane from A to B because of the random movements of its molecules. This movement of water across a semi-permeable membrane in response to a difference in solute concentration is called *osmosis*.

Strictly speaking, in osmosis, water moves in response to differences in *osmotic pressure* rather than solute concentration. The osmotic pressure (π) of a solution is directly proportional to the solute concentration (c) and can be calculated using the Van't Hoff equation:

$$\pi = icRT,$$

where R is a constant, T is the absolute temperature and π is expressed in atmospheres or bars (1 atm = 1·013 bars). For molecular solutes, $i = 1$, but for ionic solutes i is the number of ions in the formula (e.g. $i = 2$ for KCl, 3 for K_2SO_4, etc.). Thus, if c is expressed in moles/litre, ic is a measure of the total number of particles per litre which cannot cross a semi-permeable membrane. Osmotic pressure is, therefore, determined by the number of particles, rather than the molar concentration, of solute.

It might be expected that osmotic movement of water from the weaker (A) to the stronger (B) solution would continue until the concentrations of NaCl were equal at 1.5 M. However, as osmosis proceeds, a head of water develops between A and B—

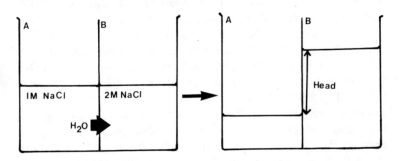

whose hydrostatic pressure opposes further water movement. At equilibrium, the osmotic flow of water ceases because the pressure of the head of water is equal to the difference in osmotic pressure between the solutions in A and B (note that this is the osmotic pressure difference at equilibrium, not at the beginning of the experiment).

A good example of this process is provided by plant cells. In a typical leaf cell, the large central vacuole contains a concentrated

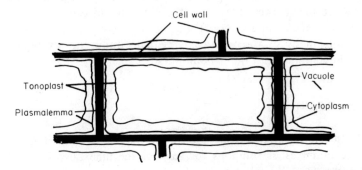

FIGURE 5.1. Schematic cross-sectional diagram of a typical mature plant cell showing the 'complex semipermeable membrane' (plasmalemma, cytoplasm and tonoplast) separating water in the cell walls from that in the vacuole. Note that cytoplasmic inclusions (nuclei and organelles) are omitted and that, in a fully turgid cell, the plasmalemma would be pressed against the cell walls.

solution of organic and inorganic solutes at an osmotic pressure of about 16 bars. The vacuolar contents are separated from the very dilute aqueous solution (osmotic pressure $\simeq 0$) bathing the outside of the cell, by the tonoplast membrane, the cell cytoplasm and the plasmalemma, which together act as a semi-permeable membrane (Figure 5.1). Consequently, if the leaf is not transpiring rapidly, water tends to move from the external solution into the vacuole by osmosis, and the volume of the vacuole tends to increase. In a leaf cell lacking cell walls, this flow of water would continue until either the cell burst or the difference in osmotic pressure was removed by dilution of the vacuolar sap. However, in a leaf, cell volume is limited by cell walls, and only a relatively small inflow of water can be accommodated by the elasticity of these walls. Consequently, hydrostatic pressure (turgor pressure) develops in the vacuole, pressing the cytoplasm against the inner surface of the cell walls. When the turgor pressure, which opposes the further entry of water, reaches about 12 bars it equals the osmotic pressure difference (reduced to about 12 bars by dilution of the vacuole) which is driving water inwards. Osmotic water flow then ceases and the cell is at maximum turgor pressure.

It may be helpful to think of the plant cell as a balloon, whose skin is the compound semi-permeable membrane of tonoplast, cytoplasm and plasmalemma, being blown up inside a rigid box, the cell walls. When the balloon fills the internal volume of the box, its volume can increase no further; the air pressure in the balloon then increases (development of turgor pressure) until it equals the maximum pressure the 'blower' can exert (the difference in osmotic pressure between vacuole and exterior). Further inflation is then impossible (osmotic inflow of water ceases). Note that without the box, inflation would continue until the balloon burst.

EXERCISES

(1) In solutions, both solute and solvent may be gas, liquid or solid. Give examples of solutions for each of the nine possible combinations, e.g. solid in liquid, NaCl in water; gas in gas, O_2 in N_2, etc.

(2) (a) Calculate the molecular weights of the following compounds: Sodium chloride $NaCl$; Methane CH_4; Ammonium sulphate $(NH_4)_2SO_4$; Ethyl alcohol C_2H_5OH; Maneb (a fungicide) $C_4H_6MnN_2S_4$.

(b) Calculate the mass of solute in 1 litre of the following aqueous solutions: 2 M sodium chloride, 0·5 M glucose; 0·01 M urea; 0·15 M ammonium nitrate, NH_4NO_3.

(c) Calculate the mass of potassium ions in the following solutions: 1 litre of 0·05 M KCl; 150 ml of 0·001 M K_2SO_4.

(3) Calculate the phosphorus content of:
   (a) a 30 g sample of pure monocalcium phosphate, $Ca(H_2PO_4)_2$;
   (b) 0·2 moles of monocalcium phosphate.

(4) Calculate the osmotic pressure (in atm) of the following solutions at 20°C (where R = 0·082 litre atm deg^{-1} $mole^{-1}$); 0·05 M glucose, 0·1 M NaCl, 0·2 M KCl, 1 M NH_4NO_3.

Are these solutions *isotonic* (have the same osmotic pressure) *hypotonic* (lower o.p.) or *hypertonic* (higher o.p.) compared with 1 M urea?

(5) Animal cells differ from plant cells in having no cell walls. Why do they not burst due to uncontrolled osmotic inflow of water?

FURTHER READING

1. PAULING L. (1970) *General Chemistry*, 3rd edition, Ch. 13. W.H. Freeman & Co., New York.
2. ROSSOTTI H. (1975) *Introducing Chemistry*, Ch. 14. Penguin Books, Middlesex. In more advanced treatments of osmosis, the flow of water is considered to be driven by differences in water potential rather than osmotic pressure. The meaning of water potential, and its use in interpreting the water relations of plant cells, is considered in:
3. FITTER A.H. & HAY R.K.M. (1981) *Environmental Physiology of Plants*, Ch. 4. (In preparation.) Academic Press, London.
4. SUTCLIFFE J. (1979) *Plants and Water*, 2nd edition. Studies in Biology No. 14. Edward Arnold, London.
Osmoregulation and semi-permeable membranes in animal tissues are discussed in:
5. LOCKWOOD A.P.M. (1971) *The Membranes of Animal Cells*. Studies in Biology No. 27. Edward Arnold, London.

CHAPTER 6
THE GAS STATE

The most important process in plant growth is the diffusion of carbon dioxide gas through stomata to the leaf mesophyll where it can be incorporated into carbohydrate molecules by photosynthesis. Since all primary (plant) and secondary (animal) production is derived from this process, it is important for environmental scientists to understand the basic physics and chemistry of gases. Knowledge of the properties of gases is also useful in understanding the properties of compressed fuel gases, the operation of the internal combustion engine, the distribution of gaseous pollutants and in most branches of meteorology.

6.1 THE GAS LAWS

Gases differ from liquids and solids in their large volume responses to changes in environmental conditions, particularly pressure and temperature. For example an increase in pressure from 1 to 2 bars causes a 50% reduction in the volume of a sample of air compared with less than 0·01% for liquid water. When water is heated from 0 to 100°C, its volume increases by only 2% compared with 36·6% for air over the same temperature range. These large changes in gas volume in response to changes in pressure and temperature can be predicted by the Gas Laws.

(a) Boyle's Law

This states that the volume (V) of a fixed mass of gas is inversely proportional to the pressure (p) if the temperature remains constant, i.e.

$$V \propto \frac{1}{p} \quad \text{or} \quad V = \frac{k_b}{p} \quad \text{or} \quad pV = k_b,$$

where k_b is a constant. In simple terms, this law states that the volume of a sample of gas decreases regularly as the pressure increases, and vice versa.

If we begin with a mass of gas at pressure p_1 and volume V_1 and then increase the pressure to p_2, the volume will fall to V_2. Therefore,

$$p_1 V_1 = k_b \text{ and } p_2 V_2 = k_b,$$

consequently,

$$p_1 V_1 = p_2 V_2 \text{ if the temperature remains constant.}$$

This last form of Boyle's Law is the most useful since it allows us to predict the result of a change in pressure without the need to evaluate the constant k_b.

(b) Charles' Law

This states that the volume (V) of a fixed mass of gas is directly proportional to the *absolute* temperature (T), if the pressure remains constant, i.e.

$$V \propto T \text{ or } V = k_c T \text{ or } \frac{V}{T} = k_c,$$

where k_c is a constant. Here the volume of a gas increases regularly as the temperature rises.

As for Boyle's Law, the most useful form of Charles' Law is

$$\frac{V_1}{T_1} = \frac{V_2}{T_2} \text{ if the pressure remains constant.}$$

(c) The General Gas Equation

These two laws do not have wide practical application, because temperature must be kept constant for Boyle's Law and pressure constant for Charles' Law. Changes in volume in real life normally involve alterations in both pressure and temperature and it is, therefore, necessary to combine the two laws to give the more useful *General Gas Equation* (G.G.E.).

The combination of the laws can be explained as follows. If we have a sample of gas of volume V_1 at a pressure p_1 and temperature T_1 then by Boyle's Law, its volume at p_2 and T_1 will be:

$$\frac{p_1 V_1}{p_2}.$$

If we now increase the temperature of the gas from T_1 to T_2 without changing the pressure, which remains at p_2, then, by Charles' Law, the volume, V_2, at the end of the second change can be calculated as follows:

$$\frac{V_2}{T_2} = \frac{\dfrac{p_1 V_1}{p_2}}{T_1}.$$

On rearrangement, this gives the most useful form of the G.G.E.:

$$\frac{p_1 V_1}{T_1} = \frac{p_2 V_2}{T_2}$$

which can be used to evaluate the change in volume of a gas sample (V_1 to V_2) which will accompany a change in pressure from p_1 to p_2 *and* a change in temperature from T_1 to T_2 (whether these changes occur in one or a number of steps). Note that if the temperature remains constant ($T_1 = T_2$) the G.G.E. reverts to Boyle's Law and if pressure remains constant ($p_1 = p_2$) it becomes Charles' Law.

In calculations using the gas laws it is useful to have reference values of temperature and pressure at which a sample of gas has a known volume. The most widely used reference values are *S.T.P.* (Standard Temperature and Pressure), i.e. 273°A (0°C), and atmospheric pressure (1 atm, 1.013 bar or 760 mmHg, depending on the units used; note that since the two units of pressure, the atmosphere and the bar, are very similar in magnitude, they tend to be used interchangeably although the bar is the more correct scientific unit).

6.2 THE LIQUEFACTION OF GASES

As it has been presented, Charles' Law appears to describe the volume changes of gases at all temperatures from 0°A upwards. However, as a gas is cooled, its molecules move more and more slowly until they finally condense to give a liquid at a temperature normally well above 0°A. Consequently, Charles' Law can be applied to a substance only at temperatures above its boiling point (at the fixed pressure used).

Increase of pressure also causes gases to condense to liquids by forcing the molecules together. For example, we can keep liquid propane under pressure in a cylinder at laboratory temperatures (15–25°C) which are far higher than the boiling point of propane (−42°C). Thus Boyle's Law is applicable only over a limited range of pressures.

The liquefaction of gases by pressure is an important feature of the petroleum fuel industry. In many parts of the world the hydrocarbon gases, methane (b. pt. $-164°C$), ethane ($-89°C$), propane ($-42°C$) and butane ($-0·5°C$), occur in huge quantities either in natural gas fields or in oil wells. These fuels must be liquefied for efficient transport and distribution but this cannot be achieved economically by costly, long-term refrigeration. Instead, the gases are liquefied by pressure and stored in strong pressurized containers for example, in 20 kg laboratory cylinders of liquid propane. No energy is needed to maintain the fuel as a liquid and when a flow of fuel gas is required, it can be supplied by opening the valve in the cylinder thus reducing the pressure and permitting the liquid to vaporize.

6.3 METHODS OF EXPRESSING THE COMPOSITION OF GAS MIXTURES

The composition of a gas mixture is commonly expressed in terms of the *percentage* of the total *volume* occupied by each gas. Thus the composition of dry air is often quoted as 78 % nitrogen (N_2), 21 % oxygen (O_2), 0·93 % argon (Ar), 0·03 % carbon dioxide (CO_2) (variable), *by volume*, the remaining volume being occupied by small quantities of the other inert gases (He, Ne, Kr), nitrogen oxides and methane. Minor constituents of air (e.g. pollutants like SO_2 and nitrogen oxides) are normally expressed as p.p.m. (volumes per million volumes—see section 5.1).

The volume percentage composition of a gas mixture is very useful because it also gives information on the ratio of the number of moles of each constituent gas. This is because, by *Avogadro's Principle*:

Equal volumes of different gases contain the same number of moles (or molecules) if they are at the same temperature and pressure. For example, it has been found that 1 mole of any gas at STP has a volume of 22.4 l.

From this we can calculate that a 22·4 l sample of air at STP contains 0·78 mole of N_2, 0·21 mole of O_2 and 0·0093 mole of Ar.

Avogadro's Principle can also be written as a form of the General Gas Equation:

$$\frac{pV}{T} = nR$$

or
$$V = \frac{nRT}{p},$$

where n is the number of moles in the gas sample and R is a constant, (the Gas Constant, which also appears in the Van't Hoff Equation section 5.4), i.e. the volume of a gas sample at constant pressure and temperature is proportional to the number of moles of gas in the sample. This form allows us to explain another widely used method for expressing the composition of a gas mixture—*partial pressures*. For example, in a given volume of air (V) contained in a bottle, the total number of moles present (n) is equal to the sum of the number of moles of the constituent gases:

$$n = n_{N_2} + n_{O_2} + n_{Ar} \text{ etc.}$$

For a given temperature, T, we can multiply each side of this equation by the constant $\frac{RT}{V}$, giving:

$$\left(\frac{RT}{V}\right)n = \left(\frac{RT}{V}\right)n_{N_2} + \left(\frac{RT}{V}\right)n_{O_2} + \left(\frac{RT}{V}\right)n_{Ar}.$$

By the G.G.E., the left-hand side of this equation, $\frac{RT}{V}n$, is now equal to p, the pressure exerted by the volume of air against atmospheric pressure (1 atm at STP—note that if a volume of gas did not exert a pressure against atmospheric pressure, it would collapse). Similarly, $\frac{RT}{V}n_{N_2}$ is equal to the *partial pressure* of nitrogen (p_{N_2}), that is, the pressure it would exert if it *alone* filled the bottle and the other constituents were absent. Similarly $\frac{RT}{V}n_{O_2} = p_{O_2}$, the partial pressure of oxygen, i.e.

$$p = p_{N_2} + p_{O_2} + p_{Ar}.$$

We have already seen that 1 'mole' of air occupies 22·4 l, exerts a pressure of 1 atm and contains 0·78 mole of N_2, 0·21 mole of O_2 and 0·0093 mole or Ar. Consequently, since

$$p = \frac{RT}{V}n = 1, \quad \frac{RT}{V} = 1$$

and, therefore, the partial pressures of nitrogen, oxygen and argon in dry air at STP are

$$p_{N_2} = 1 \times 0 \cdot 78 = 0 \cdot 78 \text{ atm}$$
$$p_{O_2} = 1 \times 0 \cdot 21 = 0 \cdot 21 \text{ atm}$$
$$p_{Ar} = 1 \times 0 \cdot 0093 = 0 \cdot 0093 \text{ atm.}$$

Note that the ratio of partial pressures of the different components is the same as the ratio of the number of moles of the components.

So far we have discussed the composition of dry air only. However, since the moisture content of the air is one of the most important environmental factors controlling plant growth, it is necessary to understand the conventional methods of expressing water vapour content.

The simplest expression employed is the *Specific Humidity* which is the mass of water vapour contained in unit mass of *moist* air (g g^{-1} or kg^{-1}). The maximum specific humidity of air, at which it is saturated with moisture, varies widely with temperature (Table 6.1) and air pressure.

TABLE 6.1. Quantities of water vapour required to saturate air at 1 atm pressure and various temperatures (g kg^{-1} of moist air).

Temperature (°C)	g kg^{-1}
0	3·8
5	5·4
10	7·7
15	10·7
20	14·7
25	20·0
30	26·9
35	35·8
40	47·3

Because of the great variability of air moisture content, another expression *Relative Humidity* is more commonly used. The relative humidity (RH) of air is calculated using the following formula:

$$\text{RH}(\%) = \frac{\text{water vapour content of the air (g } kg^{-1} \text{ air)}}{\text{water vapour content of saturated air at the same temperature and pressure (g } kg^{-1} \text{ air)}} \times \frac{100}{1}.$$

Consequently, the RH of saturated air is always 100%, whatever the pressure or temperature, and any water vapour content lower than the saturation value gives an RH value less than 100%. For example, using Table 6.1 we can calculate that the RH of air containing 5·4 g kg^{-1} water vapour falls from 100% at 5°C, through 50% at 15°C and 27% at 25°C to 15% at 35°C (and 1 atm air pressure).

For reference purposes, some important properties of oxygen and carbon dioxide in aqueous solution are given in Table 6.2.

TABLE 6.2. Gas solubilities and diffusion coefficients in water.

	10°C	20°C	30°C
Solubility of O_2 (mole l^{-1})	0·00035	0·00028	0·00024
Solubility of CO_2 (mole l^{-1})	0·0514	0·0365	0·0267

Diffusion coefficient of O_2 in water at 25°C $= 2·9 \times 10^{-5}$ cm^2 s^{-1}
Diffusion coefficient of CO_2 in water at 20°C $= 1·7 \times 10^{-5}$ cm^2 s^{-1}
(The diffusion of O_2 and CO_2 is much faster through air than through water as shown by their D values in air: $1·8 \times 10^{-1}$ cm^2 s^{-1} for O_2 and $1·4 \times 10^{-1}$ cm^2 s^{-1} for CO_2, both values measured at 0°C.)

EXERCISE

(1) Calculate the water vapour content of air at 1 atm pressure and the following temperature and RH values: 95% RH at 15°C; 22% at 35°C; 60% at 5°C.

FURTHER READING

The fundamental physics and physical chemistry of gases are treated at a more advanced level in:
1. MONTEITH J.L. (1973) *Principles of Environmental Physics*, Ch. 2. Edward Arnold, London.
The diffusion of gases (with special reference to photosynthesis and transpiration in plants) is discussed fully in:
2. MEIDNER H. & MANSFIELD T.A. (1968) *Physiology of Stomata*, Ch. 3. McGraw-Hill, London and New York.
The properties of the earth's atmosphere are considered at length in:
3. McINTOSH D.H. & THOM A.S. (1969) *Essentials of Meteorology*, Ch. 2. Wykeham.

CHAPTER 7

CHEMICAL REACTIONS

A chemical reaction is a process which transforms one or more substances into different chemical substances. A rather dramatic example, often demonstrated in the laboratory, is the reaction of sodium metal with water; this reaction, which transforms sodium metal and water into sodium ions, hydroxide ions and hydrogen gas, also produces a great deal of heat, causing the unreacted metal to melt and rush around the surface of the water until the reaction is complete. Chemical reactions should not be confused with purely *physical* processes, like changes of state (e.g. the melting of sodium), in which the chemical nature of the substance remains unaltered.

The study of chemical reactions is normally the most important single aspect of an elementary course in chemistry. However, in agriculture and ecology, we are generally more interested in the properties and uses of inorganic chemicals and, therefore, only one chapter of Part 1 of this text is devoted specifically to reactions. In Part 2, we shall consider organic reactions in much more detail since they are the foundation of biochemistry.

7.1 CHEMICAL REACTIONS AND EQUATIONS

The chemical reaction between sodium and water can be described concisely by the chemical equation:

$$2Na + 2H_2O \rightarrow 2NaOH + H_2\uparrow$$

which is the conventional shorthand expression for:

'2 moles of sodium metal react with 2 moles of water to give 2 moles of sodium hydroxide and 1 mole of hydrogen'.

Note: (a) Since atoms are neither created nor destroyed in chemical reactions, the number of atoms of each element must be the same on each side of the equation, i.e. the equation must balance.
(b) It is conventional to avoid fractions of moles in equations. Thus the equation above is preferable to the form:

$$Na + H_2O \rightarrow NaOH + \tfrac{1}{2}H_2\uparrow$$

(c) For reactions occurring in aqueous solution, an upward-directed arrow (e.g. $H_2\uparrow$) indicates that a product (here H_2) is lost from solution as a gas; similarly, a downward-directed arrow indicates the loss of a product from solution as an insoluble solid precipitate.

Since chemical reactions result from the collision of the reacting particles (atoms, ions or molecules), the faster the particles are moving, the more often they will collide with one another and react. Consequently, solid state reactions are very slow, and of little importance, because the particles are moving slowly. Most reactions of importance in biology and agricultural science occur in the liquid state and particularly in aqueous solution (cytoplasm, blood, soil solution) where the particles are moving much more rapidly. Many industrial reactions, like the production of fertilizer ammonia, are carried out in the gaseous state.

We can separate chemical reactions into a number of different classes, although it should be pointed out that a given reaction may fall into more than one class. For example:

(a) *Decomposition Reactions* involving the breakdown of a single compound into a number of simpler substances, e.g. the thermal decomposition of limestone, in lime kilns, to give quicklime (calcium oxide):

$$CaCO_3 \rightarrow CaO + CO_2\uparrow$$

(b) *Ionic Reactions* occurring between ions in aqueous solution (without breakdown of the ions), e.g. the exchange of anions between aluminium sulphate and calcium hydroxide (both soluble) giving two insoluble products:

$$Al_2(SO_4)_3 + 3Ca(OH)_2 \rightarrow 2Al(OH)_3\downarrow + 3CaSO_4\downarrow$$

(c) *Hydrolysis Reactions* involving the splitting of water molecules. For example, the hydrolysis of polysaccharides (section 13.5), lipids (13.6) and proteins (14.2) involve the splitting of H_2O into OH and H, which are attached to the products.

(d) *Polymerization and Condensation Reactions* linking together many similar molecules to give large macromolecules. For example the synthesis of polyethylene and polypropylene (12.2), polysaccharides (13.5) and proteins (14.2).

(e) *Acid/Base Reactions* Considered fully in Chapter 8.

(f) *Redox Reactions* involving the oxidation or reduction of a substance. We shall consider redox reactions in some detail in the next two sections because of their great importance in environmental chemistry as well as their economic importance.

7.2 REDOX REACTIONS

An *oxidation* reaction occurs whenever electrons are *removed* from an atom or molecule, for example, in the ionic reaction:

$$2Fe^{3+} + Sn^{2+} \rightarrow 2Fe^{2+} + Sn^{4+}.$$

Sn^{2+} ions are oxidized to Sn^{4+} ions by Fe^{3+} (ferric) ions which are the *oxidizing agents* in this reaction. The oxidation is stressed if we write the equation as a half reaction:

$$Sn^{2+} \rightarrow Sn^{4+} + 2 \text{ electrons.}$$

A *reduction* reaction occurs whenever electrons are *added* to an atom or molecule. Reduction reactions always accompany oxidation reactions (and vice versa) because reduction reactions use the electrons liberated by oxidation reactions. Thus, in the reaction:

$$2Fe^{3+} + Sn^{2+} \rightarrow 2Fe^{2+} + Sn^{4+}$$

Fe^{3+} (ferric) ions are reduced to Fe^{2+} (ferrous) ions by Sn^{2+} ions which are the *reducing agents* in this reaction. Or more simply,

$$2Fe^{3+} + 2 \text{ electrons} \rightarrow 2Fe^{2+},$$

where the electrons originate from the oxidation of Sn^{2+} ions. In more general terms, any redox reaction can be written as follows:

$$A_{oxidized} + B_{reduced} \rightarrow A_{reduced} + B_{oxidized},$$

where $A_{oxidized}$, in changing to $A_{reduced}$, is acting as an oxidizing agent, and $B_{reduced}$, in changing to $B_{oxidized}$, is acting as a reducing agent.

This fundamental explanation of redox reactions in terms of electron exchange can be extended to demonstrate that:

(i) reactions in which oxygen is added to, or hydrogen removed from, a substance are oxidation reactions; and

(ii) reactions in which oxygen is removed from, or hydrogen added to, a substance are reduction reactions.

As a simple example, the transformation of acetaldehyde (ethanal) to acetic acid (ethanoic acid) by potassium permanganate,

$$\begin{array}{ccc} & H \ \ O & & H \ \ O \\ & | \ \ \| & & | \ \ \| \\ H-C-C-H & \rightarrow & H-C-C-O-H \\ & | & & | \\ & H & & H \\ \text{ethanal} & & \text{ethanoic acid} \end{array}$$ (section 13.4),

is an oxidation reaction. The added oxygen atom is highly electro-negative and, therefore, tends to draw electrons away from the rest of the molecule,

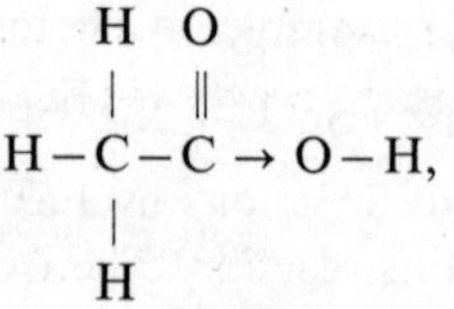

$$
\begin{array}{cc}
\text{H} & \text{O} \\
| & \| \\
\text{H}-\text{C}-\text{C} & \rightarrow \text{O}-\text{H}, \\
| & \\
\text{H} &
\end{array}
$$

giving a partial *removal* of electrons from the remainder of the molecule. Thus the molecule has been oxidized.

Important Redox Reactions

(a) Redox reactions involving iron ions are very important in soil chemistry. Under aerobic conditions, the oxidation reaction,

$$Fe^{2+} \rightarrow Fe^{3+} + le,$$

is favoured and since the ferric iron is red, well-aerated soils tend to be red or reddish-brown, depending upon the organic matter content of the soil. However, under anaerobic conditions in poorly drained soils, soil microbial respiration uses ferric ions (instead of O_2) as terminal electron acceptors:

$$Fe^{3+} + le \rightarrow Fe^{2+}$$

giving bluish-grey ferrous ions. Consequently, waterlogged soils tend to be grey whereas intermittently waterlogged soils are mottled red and grey.

(b) the reduction of iron oxide in the blast furnace:

$$Fe_3O_4 + 4C \rightarrow 3Fe + 4CO \text{ (section 4.3)}.$$

(c) Corrosion, Combustion and the Synthesis of Ammonia which are each considered in more detail in subsequent sections.

7.3 THE CORROSION OF METALS

The deterioration of metals by corrosion (rusting, etc.) is a major problem in industry and agriculture, costing many hundreds of millions of pounds each year in the UK alone. In drier climates and in areas of less severe air pollution, corrosion reactions are not so widespread but some problems are common throughout the world, for example:

(a) Rapid corrosion of the exhaust systems of motor vehicles, accelerated by the high temperatures of exhaust gases;

(b) Rusting of unprotected metal surfaces exposed to moist air (for example, the corrosion of the mouldboards of ploughs can give rise to increased friction between plough and soil);

(c) The deterioration of metals due to reactions with acids (aerial pollutants such as SO_2, spillage of battery acid, etc.).

Corrosion Reactions

As we saw in section 4.3, most metals are highly reactive and occur in nature as oxide or sulphide ores. These ores must be reduced to give the pure metal for industrial use. For example, zinc blende (ZnS), a major zinc ore, is first roasted with oxygen to give the oxide which is then reduced by heating with coke:

$$2ZnS + 3O_2 \rightarrow 2ZnO + 2SO_2$$
$$ZnO + C \rightarrow Zn + CO.$$

However, due to their reactivity, these reduced metals tend to revert to oxides and other compounds of the metal, resulting in a deterioration of the useful mechanical properties of the metal. The reactions involved in corrosion are, therefore, oxidation reactions, reversing the reduction reactions involved in their preparation. We shall examine three common types of corrosion reaction.

1. FORMATION OF OXIDE FILMS

All metals react with oxygen in the atmosphere to give a thin film of metal oxide covering the outer surface of the metal, e,g.

$$2Cu + O_2 \rightarrow 2CuO.$$

In the case of the most useful metals for construction (e.g. Al, Ni, Zn, Fe, Cu), this oxide film is very valuable because it protects the bulk of the metal from further oxidation. Oxide films are easily demonstrated by rubbing the metal surface with sandpaper to expose bright unoxidized metal.

In other, less useful metals (e.g. K, Na, Mg, Ca), the metal oxide film tends to crack and peel off, exposing fresh metal to the atmosphere. Corrosion can, therefore, penetrate deeply into the metal and damage its mechanical properties. As we shall see below, the oxide film in iron and its alloys does not protect the metal from electrochemical corrosion, and it may be disrupted by prolonged heating.

However, both of these problems can be overcome by alloying steels with Cr (13–20%), giving stainless steels whose chromium oxide film gives more complete protection from corrosion.

2. ELECTROCHEMICAL CORROSION

If pieces of two different metals are placed in contact in aqueous solution, electrons tend to flow from one metal to the other. We can predict which metal will release electrons and which will accept electrons using the *Electrochemical Series*:

$$Na\ Mg\ Al\ Ti\ Zn\ Fe\ Ni\ Sn\ Pb\ Cu\ Ag\ Pt\ Au,$$

which has been drawn up as a result of many experimental observations.

In general, electrons flow from a metal earlier in the series to a metal later in the series; in other words, the earlier metal will be oxidized. For example, if Fe and Cu are the two metals, Fe, being earlier in the series, will supply electrons to Cu by the reaction:

$$Fe \rightarrow Fe^{3+} + 3e$$

and the electrons flowing into the copper metal combine with hydronium ions in the bathing medium to give hydrogen gas:

$$2H_3O^+ + 2e \rightarrow 2H_2O + H_2\uparrow.$$

If, instead, we have a single piece of iron containing copper as an impurity (or attached to another piece of iron by copper rivets), then, if the metal remains moist, electrons will flow from iron to copper as before. The ferric ions produced will combine with hydroxyl ions in the bathing medium to give insoluble ferric hydroxide:

$$Fe^{3+} + 3OH^- \rightarrow Fe(OH)_3\downarrow.$$

This ferric hydroxide precipitate reacts further with oxygen and water to give the complex mixture of hydrated iron oxides (e.g. $Fe_2O_3.H_2O$) known as *rust*, which flakes off the metal surface, exposing unreacted metal. Since the copper metal is not consumed in this process, the rusting of iron will continue, as long as it is exposed to water and oxygen, until it is entirely degraded to rust.

Rusting is the most familiar form of electrochemical corrosion because of the widespread use of ferrous metals (section 4.3) and also because the iron and steel used in construction are rarely free of metallic contamination. However, electrochemical corrosion occurs whenever two metals are in contact and care must therefore be exercised if a piece of equipment is to be constructed from more than one metal or alloy.

The Prevention of Electrochemical Corrosion

The deterioration of metals by electrochemical corrosion can be prevented in two main ways.

Protective Barriers

Since both water and oxygen are required for the chemical reactions of rusting, the simplest method of preventing the corrosion of ferrous metals is to exclude both water and air from the metal surface. Long-term corrosion control can be achieved by coating the surface of metals with water-repellant paint or plastic, but layers of grease or oil can give shorter-term protection. As noted earlier, the chromium oxide surface films on stainless steels are also extremely effective.

Cathodic Protection

This method can be illustrated by galvanized iron, which is covered by a thin surface layer of zinc. If the zinc layer is damaged, allowing water to come into contact with both Zn and Fe, electrochemical corrosion will begin. However, since Zn is before Fe in the electro-chemical series, it is the Zn that will be oxidized and corrosion of the iron structure will not begin until the Zn layer has disappeared. This method gives protection for at least a few years under most environmental conditions.

3. CORROSION BY ACIDS

If vehicle batteries are overfilled, sulphuric acid may 'slop' out of the battery, causing corrosion of the battery support and the bodywork of the vehicle by the following oxidation reaction:

$$Fe + H_2SO_4 \rightarrow FeSO_4 + H_2\uparrow$$

i.e.
$$Fe \rightarrow Fe^{2+} + 2e.$$

Similar corrosion reactions occur when metals are exposed to air containing high levels of acidic pollutants (SO_2, nitrogen oxides, HF, HCl, etc.).

7.4 REACTIONS AND ENERGY

As well as leading to the formation of new chemical substances, chemical reactions normally involve the absorption or release of energy, usually in the form of heat (as we saw in the reaction of sodium metal with water). We can, therefore, class reactions as:

(a) *Exergonic*—overall release of energy. *Exothermic* if the energy released is in the form of heat; or

(b) *Endergonic*—overall absorption of energy. *Endothermic* if in the form of heat.

In biology, the most important endergonic reaction (or group of reactions) is Photosynthesis, in which the energy absorbed for the reaction is in the form of light. Overall:

$$6CO_2 + 6H_2O + \underset{\text{energy}}{\text{Light}} \rightarrow \underset{\text{carbohydrate}}{C_6H_{12}O_6} + 6O_2.$$

The absorbed energy is stored in the carbohydrate as chemical potential energy but it can be released by Respiration, an exergonic reaction (or series of reactions),

$$C_6H_{12}O_6 + 6O_2 \rightarrow 6CO_2 + 6H_2O + \text{Energy} \begin{cases} 673 \text{ Kcal mole}^{-1} \\ \text{or} \\ 2.8 \times 10^6 \text{ J mole}^{-1} \\ \text{for glucose} \end{cases},$$

whenever energy is required for the growth or metabolism of the plant. When the plant is subsequently eaten by a herbivore, its chemical potential energy will be used for the growth and metabolism of the animal. Thus the energy 'fixed' in photosynthesis is passed down food chains, with some loss as heat at each stage. Overall, virtually all the energy required by plants, animals, man and micro-organisms originates from photosynthesis.

Photosynthesis and respiration illustrate the important generalization that a reaction which is exergonic in one direction must be endergonic in the other. (This is a necessary consequence of the Law of Conservation of Energy, i.e. that energy cannot be created or destroyed.)

A simple exergonic reaction:

$$A + B \rightarrow C + D + \text{Energy}$$

can be illustrated by Figure 7.1a which stresses the fact that the reactants (A and B) have a higher energy content than the products (C and D). A clear example is the reaction of sodium metal with water:

$$2Na + 2H_2O \rightarrow 2NaOH + H_2\uparrow + Energy \begin{cases} 112 \text{ Kcal mole}^{-1} \\ \text{or} \\ 4\cdot7 \times 10^5 \text{ J mole}^{-1} \end{cases},$$

where the energy released is in the form of heat (melting the metal) and mechanical work (causing the molten metal to rush around the surface of the water).

In a similar way, the endergonic reaction:

$$E + F + Energy \rightarrow G + H$$

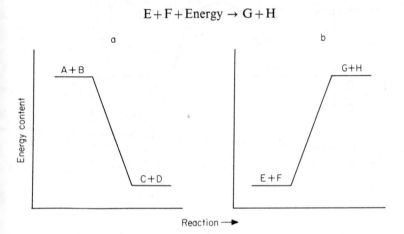

FIGURE 7.1. Energy changes during (a) a simple exergonic reaction, and (b) a simple endergonic reaction.

can be illustrated by Figure 7.1b where the reactants (E and F) have a lower energy content than the products (G and H). Instead of 'running downhill', releasing energy, an endergonic reaction proceeds 'uphill' and energy must be supplied. For example, the electrolysis of water,

$$2H_2O + Energy \rightarrow 2H_2\uparrow + O_2\uparrow$$
(approx. 68 kcal mole^{-1} or $2\cdot8 \times 10^5$ J mole^{-1})

occurs only because electrical energy is supplied.

These diagrams illustrate the simplest situations; however, in most reactions we must also consider the *activation energy* required, as shown in Figure 7.2. Here the reactants must be supplied with enough energy to push them 'uphill' over the activation energy barrier before they can react. Once they have crossed the barrier, they may 'run freely downhill', returning the activation energy and also releasing more energy, normally as the *Heat of Reaction*. This reaction is exergonic because, although activation energy is supplied, there is an overall release of energy.

A familiar example of an exothermic reaction requiring activation energy is provided by the combustion of solid fuels. Because coal and wood have higher energy contents than carbon dioxide and water

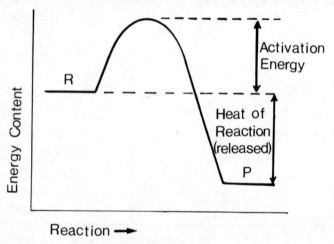

FIGURE 7.2. Energy changes during an exothermic reaction requiring activation energy (R = reactants, P = products).

(the products of burning these fuels), the combustion of coal or wood releases heat which can be used for heating or cooking. However, solid fuels do not burn spontaneously, and we have to provide activation energy (heat) using burning paper and dry kindling. Once the fuel has gained enough energy to 'cross' the activation energy barrier, it will begin to burn and to release useful heat. The reaction will then continue until all the fuel has been consumed.

Note that activation energy barriers occur also in endergonic reactions (Figure 7.3). In this case, the energy content of the reactants must be raised above that of the products before the reaction can proceed.

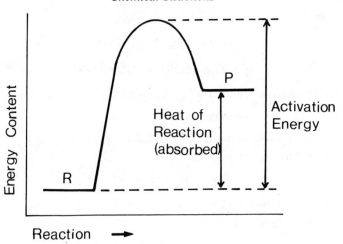

FIGURE 7.3. Energy changes during an endothermic reaction requiring activation energy. Note that, in this case, only a part of the activation energy is returned as the reaction proceeds.

7.5 CATALYSTS

A catalyst is a chemical substance which, when added to a reaction mixture, changes the rate of transformation of reactants to products, without itself being used up in the process. A catalyst alters the rate of a reaction only; it has no effect on the quantity or nature of the products obtained and it may be recovered unchanged at the end of the reaction.

Normally, catalysts are employed to speed up slow reactions. For example, in a number of gaseous reactions in the fertilizer industry, the rate of production is accelerated by finely divided metals or metal oxides.

(a) The Haber Process for the synthesis of ammonia,

$$N_2 + 3H_2 \rightarrow 2NH_3,$$

carried out at 300 atm pressure, 500–600°C over an *iron catalyst*.

(b) Synthesis of nitric acid (HNO_3),

First stage: $4NH_3 + 5O_2 \rightarrow 4NO + 6H_2O,$

carried out at 900°C over a *platinum catalyst*.
The subsequent reactions are:

$$2NO + O_2 \rightarrow 2NO_2$$

$$2NO_2 + H_2O \rightarrow HNO_3 + HNO_2$$

$$3HNO_2 \rightarrow HNO_3 + 2NO + H_2O.$$

(c) Synthesis of sulphuric acid (H_2SO_4),

First stage: $2SO_2 + O_2 \rightarrow 2SO_3$

carried out at 400–450°C over a *vanadium pentoxide catalyst*. The subsequent reaction is:

$$SO_3 + H_2O \rightarrow H_2SO_4.$$

Other examples include the hydrogenation of alkenes (section 12.2) and alkynes (12.3).

The Mechanism of Catalysis

Reactions occur as a result of the collision of the reacting molecules or atoms. For example, in the Haber Process, one molecule of N_2 and three molecules of H_2 must collide together at the same instant to give the product, ammonia. Because such multiple collisions must be very infrequent when we consider the random movement of gas molecules, the rate of reaction (moles of NH_3 produced per second) must be very low. The rate can be increased by raising the temperature, thus increasing the rate of movement of the molecules and the frequency of suitable collisions; this is the reason for the high temperatures used in the gas reactions mentioned above. Similarly, high pressures favour the reactions by forcing the molecules closer together. (The interactions between temperature and pressure are considered in more detail in the following section.)

The finely divided metal catalyst also increases the rate of reaction because it provides a large surface area on which molecules are adsorbed and assembled in position for the reaction. Thus the *shape* and great extent of the catalyst surface increase the probability that one molecule of N_2 and three molecules of H_2 will come together in a position to react.

The overall effect of a catalyst is, therefore, to lower the activation energy of the reaction, since less heat (or pressure) is required to bring the reactants into a condition to react, as shown in Figure 7.4.

Catalysis is a critical factor in many industrial reactions, speeding up production and lowering energy requirements. Catalysis is even more important in biochemistry where almost all reactions involve

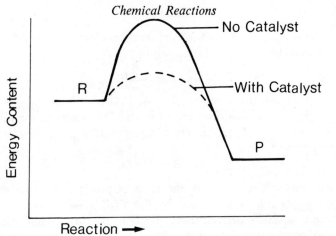

FIGURE 7.4. The effect of a catalyst on an exothermic reaction requiring activation energy. The catalyst speeds up the reaction by lowering the activation energy requirement.

protein catalysts called *enzymes*. Enzymes influence the rate of reactions by assembling the reactants on specific sites whose shapes are exactly defined by the tertiary structure of the protein molecule (section 14.2). Without enzyme catalysis, the rate of many crucial biochemical processes would be negligible.

7.6 REVERSIBLE REACTIONS

So far, we have assumed that all reactions proceed to completion and that all reactant molecules are transformed to products. However, in many cases, the reaction achieves a *stable equilibrium* when a certain amount of product has been formed but some reactant molecules are still present. In general terms, when the reaction

$$A + B \rightarrow C + D$$

reaches equilibrium, the reaction mixture contains A, B, C and D in constant proportions, and no more product can be obtained.

However, this stable equilibrium is not a static equilibrium in which all reactions have ceased. Instead, it is a dynamic equilibrium where the rate of the reaction forming the products

$$A + B \rightarrow C + D$$

is exactly equal to the rate of the reaction transforming these products back to the original reactants:

$$C + D \rightarrow A + B.$$

At equilibrium, the proportions of A, B, C and D remain constant because the rate of production of C and D is equal to the rate of consumption of C and D. This, therefore, is a *reversible* reaction, normally written as:

$$A + B \rightleftharpoons C + D.$$

For a reversible reaction, the ratio of products to reactants at equilibrium can be expressed by the *equilibrium constant* of the reaction which is derived using the *Law of Mass Action*, i.e. the rate of any chemical reaction is proportional to the (mathematical) product of the molar concentrations of the reactants. Therefore, for the reaction,

$$A + B \rightarrow C + D$$
$$\text{Rate } \alpha \text{ [A] [B]}$$
i.e. $\text{Rate}_1 = k_1[A][B]$, where k_1 is a constant.

Similarly, for the reaction,

$$C + D \rightarrow A + B$$
$$\text{Rate } \alpha \text{ [C] [D]}$$
i.e. $\text{Rate}_2 = k_2[C][D]$, where k_2 is a constant.

At equilibrium,

$$\text{Rate}_1 = \text{Rate}_2$$
$$k_1[A][B] = k_2[C][D]$$
and therefore, $$K = \frac{k_1}{k_2} = \frac{[C][D]}{[A][B]}$$

where K is the equilibrium constant for the reversible reaction.

Obviously, the higher the value of the equilibrium constant, the more product is produced by the time the reaction achieves equilibrium. Consequently, K values, which are measured under standard conditions of temperature and pressure, are very useful guides to the yield of product obtainable from a reversible reaction. For example, the Haber Process

$$N_2 + 3H_2 \rightleftharpoons 2NH_3$$

has a K value of $1 \cdot 5 \times 10^{-5}$ at $500°C$ and 1 atm pressure (if the concentrations of gases are expressed as partial pressures rather than moles l^{-1}), indicating that very little ammonia will be formed under these conditions. In the next chapter, we shall see that equilibrium constants are very useful in expressing the strengths of acids and bases.

Although equilibrium constants are valuable in the prediction of product yield under standard conditions, they cannot be used to predict how changes in conditions (temperature, pressure, etc.) will affect the equilibrium position of a reversible reaction. To do this, we must employ *Le Chatelier's Principle* which states that if the conditions of a reaction, initially at equilibrium, are changed, then the equilibrium will shift in such a direction as to tend to restore the original conditions, if such a shift is possible. We can again use the Haber Process to explain the operation of this principle.

1. INCREASE IN REACTION TEMPERATURE

The Haber Process is an exothermic reaction:

$$N_2 + 3H_2 \rightleftharpoons 2NH_3 + Heat \begin{cases} 11 \text{ kcal mole}^{-1} \\ \text{or} \\ 4 \cdot 6 \times 10^4 \text{ J mole}^{-1} \end{cases}.$$

Consequently, if we increase the temperature of the reaction after equilibrium has been achieved, then the equilibrium position (simply the ratio of reactants to products) will change in order to remove heat and restore the original conditions. This means that the decomposition of ammonia,

$$2NH_3 + Heat \rightarrow N_2 + 3H_2,$$

will be favoured since it is endothermic, and, therefore, the yield of ammonia will decrease as the temperature rises. (Note that this effect is in opposition to the general effect of raised temperature in increasing reaction rates—see above.)

2. INCREASE IN PRESSURE

The synthesis of ammonia involves a reduction in the number of moles of gas, i.e. 4 moles react to give 2, thus causing a reduction in gas pressure (section 6.3). Consequently, if we increase the pressure of the reaction after equilibrium has been achieved, the equilibrium position will change in order to reduce pressure and restore the original conditions. This favours the synthesis reaction:

$$N_2 + 3H_2 \rightarrow 2NH_3$$

and, therefore, the yield of ammonia will increase as the pressure rises.

3. REMOVAL OF PRODUCT

If ammonia is removed from the reaction mixture, then by Le Chatelier's Principle, the equilibrium will change so as to replace the lost product. Thus the continuous removal of ammonia will lead to a greatly increased total yield.

Overall, the industrial synthesis of ammonia is carried out at high pressure in the presence of a catalyst and with constant removal of the product, ammonia. Although, in theory, the yield will be reduced by the rather high temperature employed in the reaction (500–600°C), this temperature is necessary for a reasonable reaction rate.

In this chapter, a considerable amount of space has been devoted to the Haber Process for the synthesis of ammonia because it provides an excellent illustration of the use of Le Chatelier's Principle and also because it is a major part of the manufacture of nitrogen fertilizers. However, study of the Haber Process is also important in

TABLE 7.1. The energy relations of a barley crop growing in the west of Scotland.[1]

Energy inputs	GJ per ha[2]
Direct fuel use (machinery, etc)	1·42
Nitrogen fertilizer	4·88
Phosphorus fertilizer	0·26
Potassium fertilizer	0·31
Herbicide	0·05
Total[3]	6·92
Energy output	
Energy content of harvested grain[4] (Metabolizable energy)	37·8

(1) Data from J. Turnbull, quoted by M. Slessor in Chapter 1 of *Food, Agriculture and the Environment*. Eds. J. Lenihan and W.W. Fletcher (1975), Blackie, Glasgow.
(2) Where 1 GJ = 1 × 10⁹ J.
(3) Excluding a number of important inputs such as grain drying, transport, processing, etc.
(4) Grain yield = 4080 kg ha⁻¹ at 15% moisture content; metabolizable energy content of dry grain = 1·09 × 10⁷ J kg⁻¹ (from Pimental & Pimental (1979), Table 7.13 —see reading list).

explaining why the synthesis of nitrogen fertilizers requires such large inputs of fossil fuel—to maintain the high temperatures (500–600°C) and pressures (300 atm) necessary for an economic yield. Consequently, the amount of energy (in the form of oil, coal, etc.) required to produce the nitrogen fertilizer for a hectare of cereals under mechanized agriculture, can be much larger than the energy required for all other purposes (tillage, pesticides, harvesting, etc.), and it may be equivalent to a significant fraction of the energy content of the harvested crop (derived from solar radiation) (Table 7.1).

EXERCISES

(1) For practice in writing chemical equations, consult a basic chemistry or biochemistry text, write down a selection of equations omitting the number of moles and then balance the equations.

(2) Why are spontaneous reactions always exergonic reactions? Do they require activation energy?

(3) Using Le Chatelier's Principle, predict the effect of increased temperature, increased pressure and product removal on the gaseous reaction:

$$H_2 + I_2 + \text{Heat} \ (2 \cdot 6 \times 10^4 \ \text{J mole}^{-1}) \rightleftharpoons 2HI.$$

FURTHER READING

For more advanced treatment of chemical reactions, including more rigorous discussions of the thermodynamics (energy relations) of reactions consult:

1. WHITE E.H. (1970) *Chemical Background for the Biological Sciences*, 2nd Edition, Ch. 2. Prentice-Hall, New Jersey.
2. PAULING L. (1970) *General Chemistry*, 3rd Edition, Chs 10, 11, 15, 16. W.H. Freeman, New York. (More advanced treatment.)

For a discussion of the economic and practical aspects of corrosion of metals:

3. ALEXANDER W. & STREET A. (1976) *Metals in the Service of Man*, 6th Edition, Ch. 18. Penguin Books, Middlesex.

A wealth of detail about the structures and functions of enzymes can be found in:

4. MOSS D.W. (1968) *Enzymes*. Oliver and Boyd, Edinburgh.
5. WISEMAN A. & GOULD B.J. (1971) *Enzymes, their Nature and Role*. Hutchinson, London.

The various energy inputs into agricultural production (including energy for the synthesis of nitrogen fertilizers) are discussed in detail in:

6. PIMENTAL D. & PIMENTAL M. (1979) *Food, Energy and Society*. Edward Arnold, London.
7. SLESSOR M. (1975) Energy Requirements of Agriculture. In *Food, Agriculture and the Environment*, pp. 1–20. Lenihan J. & Fletcher W.W. (eds). Blackie, Glasgow.

CHAPTER 8

ACIDS AND BASES

An *Acid* is a hydrogen-containing substance which dissociates when dissolved in water, to give hydronium (H_3O^+) ions, e.g.

Sulphuric acid $H_2SO_4 + 2H_2O \rightarrow 2H_3O^+ + SO_4^{2-}$

Acetic acid $CH_3CO_2H + H_2O \rightarrow H_3O^+ + CH_3CO_2^-$.

As these examples indicate, the dissociation of an acid in water involves the transfer of a proton (H^+) from the acid to a water molecule. Consequently, acids are known as *proton donor* substances.

All acids have certain properties in common, in addition to raising the hydronium ion concentration of aqueous solutions. These include a sour or sharp taste, a tendency to 'burn' the skin, the capacity to dissolve metals and to neutralize bases (see below).

A *Base* is a substance, commonly containing the hydroxyl group or ion, which ionizes in water to give hydroxide (OH^-) ions, e.g.

potassium hydroxide $KOH \rightarrow K^+ + OH^-$

ammonia $NH_3 + H_2O \rightarrow NH_4^+ + OH^-$.

In contrast to acids, bases are *proton acceptor* substances. This characteristic is clearly shown in the ionization of ammonia, but, as we shall see in the next section, it is a feature of all bases including hydroxides like KOH. Bases have 'brackish' tastes and feel soapy. The most soluble bases (NaOH, KOH) are called *alkalis*.

In the following sections we shall investigate some of the properties of acids and bases which make them of paramount importance in biochemistry and soil science.

8.1 ACIDITY, ALKALINITY AND pH

The acidity of a solution is the concentration of hydronium ions in the solution, whereas the alkalinity of a solution can be defined as the concentration of hydroxide ions. In a given solution, these two concentration values cannot vary independently; for example, when

$[H_3O^+]$ is high, $[OH^-]$ must be low, and vice versa. The reason for this inverse relationship is as follows: The dissociation of water is a reversible reaction:

$$H_2O + H_2O \rightleftharpoons H_3O^+ + OH^-,$$

whose equilibrium constant is given by:

$$K = \frac{[H_3O^+][OH^-]}{[H_2O]^2} \qquad \text{(see section 7.6).}$$

$[H_2O]$, the molar concentration of water in water (55·6 M) is obviously a *constant* and, therefore, $[H_2O]^2$ can be absorbed into the equilibrium constant to give a new constant:

$$K_w = [H_3O^+][OH^-],$$

where K_w (the Ion Product of water) has been found by experiment to be 10^{-14} (at 25°C).

In other words, the mathematical product of the hydronium ion concentration and the hydroxide ion concentration in any aqueous solution at 25°C will be equal to 10^{-14}. For example, if the hydroxide ion concentration in a given solution is 10^{-4} M (0·0001 M), then the hydronium ion concentration will be 10^{-10} M (0·000 000 000 1 M) as shown below:

$$[H_3O^+] = \frac{K_w}{[OH^-]} = \frac{10^{-14}}{10^{-4}} = 10^{-10}.$$

Similarly, if the hydronium ion concentration rises from 10^{-10} M to 10^{-4} M, the hydroxide ion concentration will fall from 10^{-4} M to 10^{-10} M.

Using this interrelationship, we can show that solutions are acidic (contain more hydronium than hydroxide ions) when the hydronium ion concentration is higher than 10^{-7} M and alkaline when the hydroxide ion concentration exceeds 10^{-7} M. A neutral solution(neither acidic nor alkaline) contains equal concentrations of H_3O^+ and OH^-, both 10^{-7} M. (We can now see why bases are called proton acceptors. When a base dissolves in an aqueous solution, it causes an increase in hydroxide ion concentration which must be accompanied by a reduction in hydronium ion concentration according to the neutralization reaction:

$$H_3O^+ + OH^- \rightarrow 2H_2O.$$

Here each OH^- ion is accepting a proton from a hydronium ion.)

The acidity and alkalinity of some important aqueous solutions are as follows:

	Acidity $[H_3O^+]$	Alkalinity $[OH^-]$
Gastric fluid (pig)	10^{-2} M	10^{-12} M
Saliva (human)	10^{-7}	10^{-7}
Saline soil solution	down to 10^{-9}	up to 10^{-5}

In order to avoid clumsy and troublesome acidity values like 0·000 000 1 M or 10^{-7} M, it has become conventional to convert them to pH values, which fall conveniently within a 0 to 14 scale.

The pH of a solution is defined by the equation:

$$pH = -\log_{10}[H_3O^+].$$

Therefore, for pig gastric fluid:

$$pH = -\log_{10}(10^{-2}) = -(-2)$$
$$= 2.$$

Similarly, the pH of human saliva is 7 and the pH of saline soil solution can be as high as 9.

Using the definition:

$$pH = -\log_{10}[H_3O^+]$$

and the relationship:

$$K_w = [H_3O^+][OH^-] = 10^{-14},$$

we can establish the relationships between pH, $[H_3O^+]$ and $[OH^-]$ shown in Table 8.1. In using the pH scale, the following points should be stressed:

(a) as the pH value rises, $[H_3O^+]$ falls and $[OH^-]$ rises;

(b) the pH of a solution defines both the $[H_3O^+]$ *and* the $[OH^-]$ concentrations of the solution;

(c) a difference of one unit of pH indicates a tenfold difference in $[H_3O^+]$ concentration;

(d) it is not normally necessary to quote pH values to more than one decimal place (particularly since laboratory pH meters are scarcely accurate to 0·1 of a pH unit).

TABLE 8.1. The relationships between pH, H_3O^+ and OH^- in aqueous solutions.

pH	1	2	3	4	5	6	7	8	9	10	11	12	13
	Acid solutions							Alkaline solutions					
$[H_3O^+]$	10^{-1}	10^{-2}	10^{-3}	10^{-4}	10^{-5}	10^{-6}	10^{-7}	10^{-8}	10^{-9}	10^{-10}	10^{-11}	10^{-12}	10^{-13}
$[OH^-]$	10^{-13}	10^{-12}	10^{-11}	10^{-10}	10^{-9}	10^{-8}	10^{-7}	10^{-6}	10^{-5}	10^{-4}	10^{-3}	10^{-2}	10^{-1}

increasing

increasing

The Importance of pH

ENZYMES

The catalytic activity of enzymes in biochemical reactions depends upon the binding of reactant molecules to sites which are specifically shaped to accommodate these molecules only (section 7.5). However, by altering the ionization of amine base, carboxylic acid and other functional groups, variations in pH can cause changes in the distribution of charge on the surface of enzyme molecules (section 8.5) which, in turn, alter the ability of enzymes to bind reactant molecules. Consequently, each enzyme has an optimum pH range for reactant binding, and the medium in which enzyme catalysis occurs is held precisely within that pH range by means of buffers (section 8.4). For example, the digestive enzyme in human saliva operates optimally at pH 7 whereas the digestive enzymes in the human stomach (gastric fluid, pH 2) have an optimum pH value of about 2. Other aqueous media whose pH values are controlled to permit enzymes to act most efficiently include:

Bovine rumen	5·5–6·5
Vertebrate blood and lymph	7·4
Cell cytoplasm	6·9

SOILS

The pH of a soil influences the solubility of phosphate, trace elements (Fe, Mn, Zn, Cu, Co, Mo) and toxic ions (e.g. Al^{3+}) as well as controlling the activity of soil micro-organisms. For example, the fixation and mineralization of nitrogen by free-living bacteria cease at soil pH values below 4. Because of these effects, the pH values of agricultural soils are normally maintained within the range 5·5 to 7·0 to ensure adequate supplies of phosphate and trace elements, and active microbial populations, coupled with low solubilities of toxic ions. However, plant species can colonize soils whose pH values lie outside this range, if they have evolved adaptations to overcome the adverse factors associated with extremes of pH (e.g. trace element deficiencies at pH > 7).

pH is a crucial factor in many other aspects of agricultural science; to give one other example, the pH of freshly prepared silage must fall rapidly to 4·0, otherwise undesirable micro-organisms (e.g.

Clostridium spp.) will invade the mixture, giving a poor-quality or toxic silage.

8.2 THE STRENGTH OF ACIDS AND BASES

We can separate acids into two classes according to their ionization when dissolved in water, i.e.:

(a) *Strong Acids* which ionize (dissociate) completely in water, e.g.:

$$HNO_3 + H_2O \rightarrow H_3O^+ + NO_3^-$$
(Nitric acid)
$$H_2SO_4 + 2H_2O \rightarrow 2H_3O^+ + SO_4^{2-}$$
(Sulphuric acid)

These ionization reactions go to completion and are *not reversible*. In a solution of nitric acid, for example, there are H_3O^+ and NO_3^- ions only and (almost) no undissociated HNO_3.

(b) *Weak Acids* which do not ionize completely in water, e.g.:

$$CH_3CO_2H + H_2O \rightleftharpoons H_3O^+ + CH_3CO_2^- \quad \text{(Acetic Acid)}$$
$$\left. \begin{array}{l} H_3PO_4 + H_2O \rightleftharpoons H_3O^+ + H_2PO_4^- \\ H_2PO_4^- + H_2O \rightleftharpoons H_3O^+ + HPO_4^{2-} \\ HPO_4^{2-} + H_2O \rightleftharpoons H_3O^+ + PO_4^{3-} \end{array} \right\} \begin{array}{l} \text{(Phosphoric} \\ \text{acid)} \end{array}$$

These ionization reactions do not go to completion and are, therefore, reversible reactions. A solution of acetic acid contains H_3O^+ ions, $CH_3CO_2^-$ ions and undissociated CH_3CO_2H.

Consequently, a 1 M solution of a weak acid has a lower concentration of hydronium ions (i.e. a higher pH) than a 1 M solution of a strong acid. (It is important to stress at this point that the strength of an acid should not be confused with the pH of an acidic solution.)

In the same way, bases can be classed as:

(i) *Strong Bases* (the alkalis) which dissociate completely in water, e.g.:

$$KOH \rightarrow K^+ + OH^-$$

(ii) *Weak Bases* which do not dissociate completely, e.g.:

$$Ca(OH)_2 \rightleftharpoons Ca^{2+} + 2OH^-.$$

Thus, an aqueous solution of calcium hydroxide contains Ca^{2+}

ions, OH$^-$ ions and undissociated Ca(OH)$_2$.

We can express the strength of an acid (i.e. the extent of its ionization) using the equilibrium constant of its ionization. Thus for the ionization of acetic acid:

$$CH_3CO_2H + H_2O \rightleftharpoons H_3O^+ + CH_3CO_2^-$$

$$K = \frac{[H_3O^+][CH_3CO_2^-]}{[CH_3CO_2H][H_2O]} \text{ (see section 7.6)}.$$

Since [H$_2$O] is a constant, it can be absorbed into K to give a new constant,

$$K_a = \frac{[H_3O^+][CH_3CO_2^-]}{[CH_3CO_2H]}$$

and since K_a (the Dissociation Constant) is a ratio of ionized to unionized acetic acid, it can be used as an index of the strength of the acid. K_a values for a selection of acids are given in Table 8.2; the lower the value of K_a, the weaker the acid.

The dissociation constant, K_a, of an acid indicates the extent of ionization of the acid when it alone is dissolved in water. However, Table 8.2 can also be used to predict the outcome of dissolving two (or more) acids in the same solution. In all cases, the stronger acid will ionize according to its K_a value but the ionization of the weaker acid will be suppressed. Thus, in a solution containing equal concentrations of hydrochloric and acetic acids, HCl will dissociate completely to give H$_3$O$^+$ and Cl$^-$ ions whereas the acetic acid will exist almost entirely as the undissociated acid CH$_3$CO$_2$H.

TABLE 8.2. Dissociation constants of some common acids (25°C).

Dissociation	K_a	pK_a
HCl + H$_2$O $\rightarrow$ H$_3$O$^+$ + Cl$^-$ (Hydrochloric Acid)	10^7	strong acid
H$_3$PO$_4$ + H$_2$O $\rightleftharpoons$ H$_3$O$^+$ + H$_2$PO$_4^-$	7.5×10^{-3}	2.1
CH$_3$CO$_2$H + H$_2$O $\rightleftharpoons$ H$_3$O$^+$ + CH$_3$CO$_2^-$	1.8×10^{-5}	4.8
H$_2$CO$_3$ + H$_2$O $\rightleftharpoons$ H$_3$O$^+$ + HCO$_3^-$ (Carbonic Acid)	4.3×10^{-7}	6.4
H$_2$PO$_4^-$ + H$_2$O $\rightleftharpoons$ H$_3$O$^+$ + HPO$_4^{2-}$	6.2×10^{-8}	7.2
C$_6$H$_5$OH + H$_2$O $\rightleftharpoons$ H$_3$O$^+$ + C$_6$H$_5$O$^-$ (Phenol)	1.3×10^{-10}	9.9
HCO$_3^-$ + H$_2$O $\rightleftharpoons$ H$_3$O$^+$ + CO$_3^{2-}$ (Bicarbonate)	5.6×10^{-11}	10.3
HPO$_4^{2-}$ + H$_2$O $\rightleftharpoons$ H$_3$O$^+$ + PO$_4^{3-}$	2.2×10^{-13}	12.7

Another use of the dissociation constant can be explained as follows. If we imagine that the dissociation of acetic acid molecules in aqueous solution has been suppressed by a very low pH (i.e. presence of a strong acid), then as we raise the pH, the molecules begin to dissociate to give acetate ions. At a certain pH, half of the acid molecules will have dissociated to give acetate ions, i.e.

$$[CH_3CO_2H] = [CH_3CO_2^-].$$

Therefore, the equation:

$$K_a = \frac{[H_3O^+][CH_3CO_2^-]}{[CH_3CO_2H]}$$

simplifies to give

$$K_a = [H_3O^+]$$

i.e. the dissociation constant of the acid is equal to the hydronium ion concentration at which the acid is 50% ionized. Taking negative logarithms we obtain

$$-\log_{10} K_a = -\log_{10} [H_3O^+]$$

or

$$pK_a = pH$$

i.e. the pK_a of an acid is the pH at which it is 50% ionized. Some pK_a values are given in Table 8.2.

This relationship allows us to begin to predict the ionic form in which acids exist in biological fluids, soil solutions and other aqueous media. For example, since the ionization:

$$H_2PO_4^- + H_2O \rightleftharpoons H_3O^+ + HPO_4^{2-}$$

has a pK_a of 7·2, phosphoric acid added to vertebrate blood (pH 7·4) will ionize to give approximately 50% $H_2PO_4^-$ ions and 50% HPO_4^{2-} ions. HPO_4^{2-} will increasingly predominate as the pH rises above 7·4, whereas the concentration of $H_2PO_4^-$ ions is higher below pH 7·4. More complete calculations give the proportions of the four species (H_3PO_4, $H_2PO_4^-$, HPO_4^{2-}, PO_4^{3-}) shown in Figure 8.1.

The strength (degree of ionization) of bases can also be expressed using dissociation constants, e.g. for NH_3:

$$NH_3 + H_2O \rightleftharpoons NH_4^+ + OH^-$$
$$K = \frac{[NH_4^+][OH^-]}{[NH_3][H_2O]}$$

and, therefore,

$$K_b = \frac{[NH_4^+][OH^-]}{[NH_3]}$$

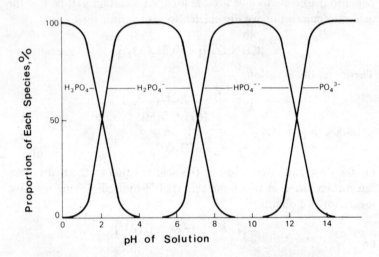

FIGURE 8.1. Proportions of phosphoric acid and its ions at different pH values. Note that over the pH range of most agricultural soils (5–7), the predominant ion in solution is $H_2PO_4^-$. (From *The Nature and Properties of Soils*, 8th edition, by N.C. Brady. © 1974 by Macmillan Publishing Co, Inc.)

Dissociation constants for a selection of amine bases are presented in Table 14.1; the lower the value of K_b, the weaker the base.

The pH of a soil is a measure of the hydronium ion concentration in the soil solution. Most of these hydronium ions originate from ionization reactions, e.g. I, where the 'colloid' in the equation represents a clay or humus particle in the soil (see sections 10.4 and 10.5 for a discussion of colloid properties). The equation

I

suggests only two ionizable hydrogens per colloid particle, for clarity, although each particle will, in practice, carry a large number.

The pH resulting from such dissociations depends upon

(a) a representative K_a value (difficult to assign a meaningful value), and

(b) the number of ionizable hydrogen atoms on the clay and humus.

Consequently, any treatment which reduces the number of ionizable hydrogen atoms attached to the soil colloids will tend to raise the pH of the soil. This is the reason for treating acidic soils with lime ($CaCO_3$, CaO or $Ca(OH)_2$); Ca^{2+} ions replace the hydrogen atoms on the colloid by the *ion exchange* reaction (II), and the resulting

II

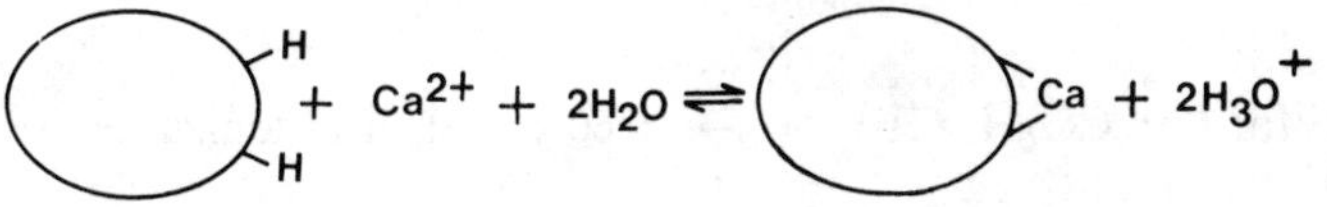

hydronium ions are neutralized by hydroxide ions, also supplied by the lime:

$$2H_3O^+ + 2OH^- \rightarrow 4H_2O.$$

The overall result of these reactions will be a rise in soil pH. Ion exchange reactions are considered in more detail in section 9.5.

8.3 EQUIVALENT WEIGHTS AND NORMALITY

Acids and bases react together by neutralization reactions, e.g.:

$$NaOH + HCl \rightarrow H_2O + NaCl$$

or more simply

$$OH^- + H_3O^+ \rightarrow 2H_2O.$$

Consequently, 1 litre of 1 M NaOH will neutralize 1 litre of 1 M HCl. However, in the reaction:

$$2NaOH + H_2SO_4 \rightarrow Na_2SO_4 + 2H_2O,$$

2 litres of 1 M NaOH are required to neutralize 1 litre of 1 M H_2SO_4, because H_2SO_4 molecules contain two ionizable hydrogen atoms.

These examples illustrate the advantage of using *gram equivalent weights* (masses of each substance yielding 1 g of protons or 17 g of hydroxide ions) rather than gram molecular weights in acid/base reactions. In formal terms:

> The gram equivalent weight of an acid is the gram molecular weight divided by the number of ionizable hydrogen atoms in the formula.

Thus for HCl, the GEW is $\dfrac{36 \cdot 5}{1} = 36 \cdot 5$ g

for H_2SO_4, GEW $= \dfrac{98}{2} = 49$ g

for CH_3CO_2H, GEW $= \dfrac{60}{1} = 60$ g (only 1 H ionizable).

> The gram equivalent weight of a base is the gram molecular weight divided by the number of hydroxide ions produced per 'molecule'.

Thus for NH_3, GEW $= \dfrac{17}{1} = 17$ g

for $Ca(OH)_2$, GEW $= \dfrac{74}{2} = 37$ g

for NaOH,   GEW $= \dfrac{40}{1} = 40$ g.

Thus 1 litre of NaOH solution, containing 1 gram equivalent weight (40 g) per litre will neutralize 1 litre of H_2SO_4 solution, containing 1 gram equivalent weight (49 g) per litre. Or, more concisely, 1 litre of 1 N NaOH will neutralize 1 litre of 1 N H_2SO_4 where a *1 N (1 Normal)* solution contains 1 gram equivalent weight/litre.

Gram equivalent weights and normality are also used for ionic solutes in ion exchange and redox reactions. In ion exchange reactions (see section 9.5), ionic charge is of paramount importance, and, therefore, the gram equivalent weight is obtained simply by dividing the gram ionic weight by the charge on the ion, e.g.

$$\text{for } H_3O^+, \text{ GEW} = \frac{19}{1} = 19 \text{ g}$$

$$Ca^{2+}, \text{ GEW} = \frac{40}{2} = 20 \text{ g}$$

$$Fe^{3+}, \text{ GEW} = \frac{56}{3} = 18 \cdot 7 \text{ g}$$

$$SO_4^{2-}, \text{ GEW} = \frac{96}{2} = 48 \text{ g.}$$

However, in redox reactions, the number of electrons *transferred* is more important than the value of the charge on the ion; consequently, for Fe^{+3}, which can accept one electron by the redox half reaction:

$$Fe^{3+} + 1e^- \rightarrow Fe^{2+},$$

$$\text{GEW} = \frac{56}{1} = 56 \text{ g.}$$

Thus for ion exchange reactions, GEW is a measure of the mass of ion providing unit electrical charge whereas, in redox reactions, it is a measure of the mass of ion exchanging unit quantity of electrons; since these definitions can result in different GEW values for the same ion (e.g. Fe^{3+} above), it is essential to use the correct procedure consistently for each application.

8.4 BUFFERS

As we noted in section 8.1, the pH of biological fluids must be controlled precisely to ensure efficient enzyme catalysis. For example, human blood is maintained within $0 \cdot 1$ of $7 \cdot 4$ since a variation of as little as $0 \cdot 2$ could prove fatal. In nature, steady pH values are usually achieved using a buffer system, based on the dissociation of a weak acid, which 'mops up' any H_3O^+ and OH^- ions produced by the metabolism.

The mechanism of buffering can be illustrated by the carbonic acid/bicarbonate ion buffer system. In solution, we have the dynamic equilibrium:

$$H_2CO_3 + H_2O \rightleftharpoons HCO_3^- + H_3O^+ \tag{8.1}$$

If we disturb this equilibrium by adding H_3O^+ ions (i.e. lowering the pH), then by Le Chatelier's Principle the equilibrium position will change so as to remove the added H_3O^+ ions. Therefore, the reaction:

$$HCO_3^- + H_3O^+ \rightarrow H_2CO_3 + H_2O \qquad (8.2)$$

will proceed until all the added H_3O^+ ions have been removed from solution and the original pH is re-established.

If instead, we add OH^- ions, they will react with H_3O^+ ions:

$$H_3O^+ + OH^- \rightarrow 2H_2O \qquad (8.3)$$

thus raising the pH. In this case, according to Le Chatelier's Principle the ionization reaction:

$$H_2CO_3 + H_2O \rightarrow HCO_3^- + H_3O^+ \qquad (8.4)$$

will be favoured until the original pH has been regained. This 'mopping-up' of H_3O^+ and OH^- ions can occur only if both elements of the system are present in solution, i.e. the anion and the un-dissociated weak acid.

The pH of the buffered solution is related to the pK_a of the appropriate weak acid as follows (using the carbonic acid/bicarbonate system as an example):

$$K_a = \frac{[H_3O^+][HCO_3^-]}{[H_2CO_3]}$$

Therefore,

$$[H_3O^+] = K_a \frac{[H_2CO_3]}{[HCO_3^-]}$$

and

$$pH = pK_a - \log \frac{[H_2CO_3]}{[HCO_3^-]}$$

$$= pK_a + \log \frac{[HCO_3^-]}{[H_2CO_3]}.$$

For example, human blood, containing 0·025 M bicarbonate and 0·00125 M carbonic acid is buffered at:

$$pH = 6 \cdot 1^* + \log\left(\frac{0 \cdot 025}{0 \cdot 00125}\right)$$

$$= 7 \cdot 4.$$

* (pK_a for dissociation in *blood*; compare Table 8.2 with Dawber J.G. & Moore A.T. *Chemistry for the Life Sciences*, 2nd Ed., Section 5.6.3. Macmillan, London.)

This buffer system is principally employed in removing hydronium ions generated by tissue metabolism, with the excess carbonic acid produced by the buffer (equation 8.2) being lost from the lungs as CO_2, as a result of the enzyme-catalyzed reaction:

$$H_2CO_3 \rightarrow H_2O + CO_2 \uparrow.$$

Thus the bicarbonate ion concentration remains very much higher than the carbonic acid concentration, thereby ensuring a steady blood pH.

8.5 AMPHOTERIC SUBSTANCES

Certain substances, notably the amino acids and proteins (section 14.2) can act either as acids (proton donors) or as bases (proton acceptors) under different circumstances. Such substances, which can contribute to the pH-buffering of living tissues, are called *amphoteric*. For example, the α-amino acid glycine has a carboxylic acid group (weak acid) at one end of the molecule and an amine group (weak base) at the other:

$$NH_2 - CH_2 - CO_2H.$$

When glycine is dissolved in water at near-neutral pH, both groups dissociate to give a *zwitterion* (charged at both ends):

$$\overset{+}{NH_3} - CH_2 - CO_2^-$$

As the pH of the solution is lowered, the ionization of the carboxylic acid group is suppressed and the molecule acts as a proton acceptor:

$$\overset{+}{NH_3} - CH_2 - CO_2^- + H_3O^+ \rightarrow \overset{+}{NH_3} - CH_2 - CO_2H + H_2O,$$

giving a positively charged ion.

In contrast, as the pH is raised, the ionization of the amine group is suppressed and the molecule acts as a proton donor:

$$\overset{+}{NH_3} - CH_2 - CO_2^- + OH^- \rightarrow NH_2 - CH_2 - CO_2^- + H_2O,$$

giving a negatively charged ion.

These changes can be detected using an electrophoresis apparatus, in which a solution of the substance under test is subjected to an electric field. At low pH values, the glycine molecule (as a positive ion) moves towards the negative pole (cathode), whereas at high pH values, the molecule (negative ion) moves towards the anode. At a definite intermediate pH value (6.1), called the *isoelectric point*, the glycine molecule does not move in either direction because the charges on each end of the molecule are the same. Since each of the 22 important α-amino acids has a different isoelectric point (due to the influence of the side-chains), electrophoresis is a useful tool in the identification of amino acids.

EXERCISES

(1) Calculate the normality of the following aqueous solutions: $109 \cdot 5$ g l^{-1} of HCl; $109 \cdot 5$ g l^{-1} of H_3PO_4; $1 \cdot 7$ g l^{-1} of NH_3; $1 \cdot 7$ g l^{-1} of aniline, $C_6H_5NH_2$.

(2) What volume of 1 M NaOH solution would be required to neutralize the following acidic solutions: 100 ml of 4 N HCl; 25 ml of $0 \cdot 33$ M H_2SO_4; 150 ml of $1 \cdot 4$ N phenol, C_6H_5OH?

(3) Use Le Chatelier's Principle to show why the addition of hydrochloric acid to an acetic acid solution causes the suppression of the ionization of the acetic acid.

(4) Why can strong acids not be used in buffer systems?

(5) Write equations to show how the following buffer systems work:

$$CH_3CO_2H/CH_3CO_2^-$$

$$H_2PO_4^-/HPO_4^{2-}$$

FURTHER READING

For more detailed treatment of acids, bases and buffers:

1. MORRIS J.G. (1974) *A Biologist's Physical Chemistry*, 2nd Edition, Chs 5 & 6. Edward Arnold, London.
2. PAULING L. (1970) *General Chemistry*, 3rd Edition, Ch. 14. W.H. Freeman, San Francisco. (More advanced.)
3. WHITE E.H. (1970) *Chemical Background for the Biological Sciences*, 2nd Edition, Ch. 2. Prentice-Hall, Inc., New Jersey.

For a basic discussion of the importance of soil pH:

4. BRADY N.C. (1974) *The Nature and Properties of Soils*, 8th Edition, Ch. 14. Macmillan, New York.

The acid/base equilibria studied in section 8.4 are also involved in the regulation of the pH and carbon dioxide content of natural waters. For further details, consult:

5. STUMM W. & MORGAN J.J. (1970) *Aquatic Chemistry*, John Wiley & Sons, Inc., New York. (Rigorous, and fairly advanced, treatment).

Using indicator (dye) molecules whose colours alter as the solution pH changes, it is possible to determine the exact quantity of acid required to neutralize a basic solution, and vice versa (section 8.3). The use of indicators in the quantitative analysis of acids and bases (acid/base titrations) is discussed fully in:

6. JOHNSTONE N.B.B. & DOWNIE T.C. (1961) *Titrimetric Analysis*. University of London Press.

CHAPTER 9

WATER

Water is the major constituent of both plants (over 90% in annuals) and animals (typically 60–70% in vertebrates). In living tissues, water is the medium for many biochemical reactions and excretion processes; inorganic nutrients, photosynthate, gases and hormones are all transported in aqueous solution; evaporation of water can control the temperature of skin or leaf; soil nutrients are available to plant roots only when dissolved in soil water. In short, water is essential for life and plays a unique role in virtually all biological processes.

In addition to this, water fulfils several other roles in agriculture. For example, most pesticides are applied to crops or livestock as aqueous solutions or suspensions, and vast quantities of water are used in irrigation and fish culture. Water also plays a central part in soil formation and erosion.

The paramount importance of water in agricultural production can be stressed by the following facts:

(a) it has been estimated that between 35 and 45 metric tonnes of water are required for the production of each kilogram of beef in the USA (this includes water for all purposes—fodder production, dipping, carcase preparation, etc., as well as the physiological requirements of the animal).

(b) Of the soil moisture absorbed by a maize crop, 98% is lost by transpiration; the remainder is retained in the plant body and only 0·2% is chemically broken down in photosynthesis.

In this chapter we shall see that although water is a very familiar substance it has unusual properties which make it uniquely suited to fulfil its many roles.

9.1 THE STRUCTURE OF LIQUID WATER

In sections 2.4 and 2.7, we saw that the water molecule is planar and angled and that, due to the high electronegativity of oxygen, there is a partial separation of charge (I) leading to the formation of hydrogen bonds between molecules (II).

100

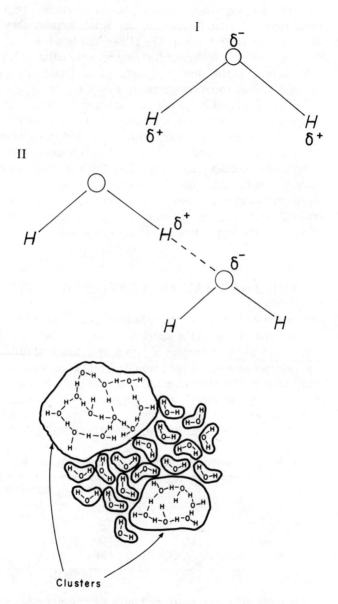

FIGURE 9.1. The 'Flickering Cluster' structure of liquid water, in which 'free' water molecules are continually exchanging with hydrogen-bonded molecules in the clusters. (From Némethy G. & Scheraga H.A. (1962) Structure of water and hydrophobic bonding in proteins. *J. Chem. Phys.*, **36**, 3382.)

Note that although these hydrogen bonds are strong compared with other intermolecular forces (van der Waals bonds), they are only about 1/24th as strong as an $O-H$ covalent bond.

If this system of hydrogen bonding between adjacent water molecules were to extend uniformly throughout liquid water, it would be solid (ice) at room temperature, with a regular crystal structure rather like Figure 4.2. Clearly, liquid water is not a solid at room temperature and, therefore, it is generally thought that it is made up of a large number of small clusters (or 'icebergs') of hydrogen-bonded water molecules separated by volumes containing free unbonded molecules only (Figure 9.1). Each of these clusters, which have a rather open, cage-like structure, is continually exchanging water molecules with other clusters and with the pool of unbonded molecules. This 'flickering cluster' structure is, therefore, dynamic, in contrast to the static structure of crystalline solids.

9.2 THE THERMAL PROPERTIES OF WATER

When a liquid is heated, the random movements of its molecules become more rapid as the temperature rises (section 3.1). However, since many of the molecules in liquid water are held firmly together by hydrogen bonds, a large amount of heat is required to overcome this intermolecular bonding and allow the molecules to move more rapidly. Consequently, the *specific heat* of liquid water (the heat required to raise the temperature of 1 g of the substance by 1°C) is very high compared with less structured liquids (Table 9.1).

TABLE 9.1. The specific heats of selected molecular substances.

	Molecular wt	Specific heat
Water, H_2O	18	4·2 $Jg^{-1}\,°C^{-1}$
Ethanol, C_2H_5OH	46	2·2
Acetone, CH_3COCH_3	58	2·1
Carbon tetrachloride, CCl_4	154	0·8

Similarly, when ice melts, or liquid water evaporates, large quantities of heat energy are needed to break the hydrogen-bonded structure so that the molecules can escape and move more freely. As a result, the temperatures of these changes of state are unusually high, compared with other molecular substances, as are the Latent Heats of Melting and Vaporization (Table 9.2); where the *Latent Heat of Melting* of a substance is the heat required to change 1 g of the substance from the solid state at its melting point to the liquid state at the same temperature; similarly for the *Latent Heat of Vaporization*, from liquid to gaseous state.

TABLE 9.2. Thermal properties of selected molecular substances.

	m.p. (°C)	b.p. (°C)	Latent heat of melting (Jg^{-1})	Latent heat of vaporization (Jg^{-1})
Water	0	100	334	2260
Ammonia	−78	−33	332	1374
Ethanol	−117	79	105	854
Acetone	−95	56	98	521
Carbon tetrachloride	−23	77	18	194

The influence of hydrogen bonding on changes of state is clearly illustrated by the melting and boiling points of the hydrides of Group 6 elements, i.e. H_2O, H_2S, H_2Se and H_2Te. If, as we might at first expect, the intermolecular bonding of these substances were similar, then we should predict that H_2O (the lightest molecule) would have the lowest m.p. and b.p. and H_2Te (the heaviest) would have the highest values. (Less energy should be required to accelerate a lighter particle.) However, as shown in Table 9.3, the relationship between molecular weight and the temperatures of melting and boiling holds only for H_2S, H_2Se and H_2Te. In H_2O, the intermolecular hydrogen bonds are so much stronger than in the other three hydrides that it has the highest m.p. and b.p. rather than the lowest. (It is worth stressing that according to its molecular weight, water should be a gas at room temperature.)

TABLE 9.3. Melting and boiling points for the hydrides of Group 6 elements.

Hydride	Molecular wt	m.p.°C	b.p.°C
H_2O	18	0	100
H_2S	34	−86	−61
H_2Se	81	−60	−42
H_2Te	130	−49	−2

Because of its high specific heat, water is a very good temperature 'buffer', absorbing or losing large amounts of heat without excessive variation in temperature. This is particularly important in warm-blooded animals whose body temperatures must be maintained close to an optimum value. It is also important in controlling the temperature of soils; the surface layers of a bare dry soil can experience enormous daily fluctuations of temperature (variations of 30–40°C are possible), causing considerable damage to plants. In moist soils, the high specific heat of water causes such fluctuations to be 'damped', giving lower maximum and higher minimum temperatures. In general, the high specific heat of water has a moderating effect on the temperatures of plants and animals, soils, lakes, the atmosphere, etc.

The high latent heats of water are also important in the control of temperature. In particular, both plants and animals remove excess heat from their surfaces by the evaporation of water. For each gram of water lost to the atmosphere by perspiration or transpiration (at 25°C), 2441 J of heat is also lost from the skin or leaf surface. On the other hand, the high latent heat of vaporization of water accounts for the serious burns or scalds caused by the exposure of skin to steam; here the condensation of each gram of steam releases 2260 J of heat (at 100°C).

In temperate agriculture, the damage to crop plants caused by severe frost can often be avoided by spraying the crop with water. As the applied water begins to freeze on the foliage of the crop, the latent heat released (334 J g^{-1}) maintains the temperature of the plant tissues above 0°C.

9.3 OTHER PROPERTIES OF WATER

Water as a Solvent

Water is an excellent solvent for three groups of biologically important solutes:

(a) *Organic solutes with which water can form hydrogen bonds*, including amino acids and low molecular weight carbohydrates, and proteins, which contain hydroxyl, amine or carboxylic acid functional groups. Water also forms colloidal dispersions (see sections 10.4 and 10.5) with larger carbohydrate and protein molecules (e.g. the cytoplasm). Similarly, water 'wets', or adheres to, solid substances such as glass, cotton and clay minerals which carry hydroxyl groups on their surfaces.

(b) *Charged ions* such as ionic salts dissolve readily because (partially charged) water molecules orientate themselves round ions in the crystal and 'pull' them into solution as highly soluble, hydrated ions (section 2.5). In the same way, water molecules become attached to fixed charges on the surfaces of plant cell walls, cell membranes and soil particles, giving tightly bound layers of water a few molecules thick.

(c) *Small molecules*, like the atmospheric gases, which can fit into holes in the open, cage-like, structure of liquid water (see section 9.1). Thus as well as being an ideal solvent for biochemical reactions, water is also a suitable medium for the transport of organic molecules (e.g. sucrose in blood and phloem), nutrient ions (e.g. nutrients from root to leaf in the xylem) and atmospheric gases (e.g. movement of O_2 to sites of respiration).

Substances which dissolve in, or bind water (groups (a) and (b)) are termed *hydrophilic*, i.e. water loving, whereas those which neither dissolve in, nor bind water are *hydrophobic*, water hating. Hydrophobic substances carry neither electrical charges nor groups forming hydrogen bonds with water; they include hydrocarbons, halogenated hydrocarbons, fats, oils, greases, etc. Many organic molecules are *amphipathic*, containing both hydrophobic and hydrophilic groups.

Volume Changes in Ice and Water

In general, when solids are heated, they expand, and this expansion continues throughout the solid and liquid states. However, as shown in Figure 9.2, water undergoes an unexpected *contraction* around its melting point; as would be predicted, ice expands with rising temperature but, at its melting point, there is a large and sudden contraction in volume ($\simeq 9\%$) followed by a more gradual contraction between 0 and 4°C, at which temperature liquid water achieves its highest density (Figure 9.2). Above 4°C, the liquid reverts to a more conventional steady expansion with increasing temperature.

The reason for this unusual behaviour is that the crystal structure of ice is a rather open, cage-like lattice (as in the 'icebergs' of liquid water). When the crystal structure breaks down on melting, individual water molecules are more mobile and can approach one another more closely, thereby occupying a smaller volume.

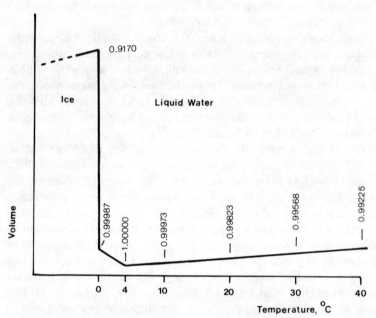

FIGURE 9.2. Changes in the volume of a fixed mass of water between 0 and 40°C (not to scale). The density of water (or ice) is given at ten-degree intervals and at 4°C, the temperature of maximum density and minimum volume.

Consequently, when a body of water (puddle, pond, lake) cools from 10°C, the cooler (denser) water falls to the bottom until the temperature of all the water has fallen to 4°C. Water cooler than this critical temperature is less dense and remains at the surface, where ice forms first. This phenomenon, which is of the greatest importance for the survival of aquatic organisms in cold climates, is a clear example of the unique relationship between the properties of water and the needs of living organisms.

Tensile Strength and Viscosity

Two other properties, tensile strength and viscosity, are highly important in the movement of water in animals, plants and soils. In particular, the unusually high tensile strength, or cohesion, of water columns in the xylem (due again to intermolecular hydrogen bonding) means that water can be drawn to the tops of tall trees by transpirational pull alone.

TABLE 9.4. The viscosity of a range of liquid substances.

	Viscosity (centipoises)	Temperature (°C)
Acetone	0·316	25
Carbon tetrachloride	0·969	20
Water	1·002	20
Ethanol	1·200	20
Glycerol	1·490	20
Mercury	1·554	20
Light machine oil	113·8	15
Heavy machine oil	660·6	15

The viscosity of a liquid is a measure of its 'stickiness' and, therefore, it controls the rate at which the liquid will pour from a spout or flow along a pipe. The viscosity of a liquid can be measured by timing the fall of a metal sphere through a column of the liquid; familiar liquids of high viscosity include treacle, honey and thick lubricating oils. In spite of its strong intermolecular hydrogen bonds, water has a moderate viscosity (when compared with other molecular liquids (Table 9.4)), permitting rapid mass flow through pipes (irrigation pipes, blood vessels, soil pores, xylem conduits, etc.) in response to modest pressure gradients. The rapid fall in the viscosity of water with rising temperature (Table 9.5) also has practical consequences; for example, tropical soils drain more rapidly than temperate.

TABLE 9.5. The influence of temperature on the viscosity of liquid water.

Temperature °C	0	5	10	15	20	25	30	35
Viscosity (centipoises)	1·787	1·516	1·306	1·138	1·002	0·890	0·798	0·719

Other Properties

The transparency of liquid water ensures a supply of light for photosynthesis to submerged plants and algae; similarly the transmission of light through cytoplasm means that energy for photosynthesis can be captured by several layers of cells in the mesophyll of the leaves of terrestrial plants. Other important properties of water include its *dissociation into ions* (already discussed in section 8.1) and its unusually high *surface tension* (see section 10.1).

9.4 THE PURIFICATION OF WATER

Water has many uses and for each use it must achieve certain standards of purity. For example, it is necessary for irrigation water to contain low concentrations of salts so as to avoid the build up of soil salinity under conditions of rapid evaporation.

For industrial and household (washing) purposes, 'soft' water, with a low dissolved salt content, is preferred. 'Hard' water, containing significant concentrations of Ca, Mg and Fe salts (particularly bicarbonates and carbonates) normally originates from limestone catchments or from boreholes receiving large quantities of leached salts. Over a period of time, calcium salts from hard water accumulate as deposits of 'fur' or 'scale' in boiler pipes and in kettles, reducing water flow and decreasing the efficiency of heating. The salts in hard water also cause the precipitation and inactivation of soaps and detergents (see section 10.2). These problems can be overcome by passing the water through an ion exchanger (section 9.5) or by the use of special synthetic detergents (section 10.2).

Rigorous controls must be exercised over the quality of piped drinking water which must be free of suspended solids, harmful micro-organisms and toxic chemicals. Normally, most of the solids and micro-organisms are removed by passing the water through sedimentation tanks and slow filter beds; the remaining organisms can then be killed by the addition of chlorine or another sterilizing agent. The exclusion of toxic chemicals can be achieved only by firm controls over the catchment area, e.g. ensuring pesticides do not drain directly into reservoirs, and by regular analysis of the purified drinking water.

In the laboratory, we require pure water containing very low concentrations of all impurities. For most purposes, water distilled using glass apparatus is sufficiently pure, but even this contains, amongst other impurities, alkali metal ions dissolved out of the glass. Ultrapure water for analytical work can be obtained by distillation using tin equipment followed by deionization, i.e. the removal of both cations and anions using an ion exchange resin (section 9.5).

However, it must be stressed that even ultrapure distilled and deionized water must contain hydronium and hydroxide ions (each at 10^{-7} M) according to the relationship:

$$K_w = [H_3O^+][OH^-] = 10^{-14}.$$

In addition, distilled water rapidly absorbs gaseous impurities (CO_2, O_2, N_2, etc.) when left open to the atmosphere.

9.5 ION EXCHANGE IN SOILS AND IN THE PURIFICATION OF WATER

Ion Exchange in Soil

Clay and humus particles carry, on their surfaces, negative charges which can bind cations, particularly H^+, NH_4^+, $*Al^{3+}$, Ca^{2+}, Mg^{2+}, and K^+. These bound cations are important in determining the pH of the soil (see section 8.2) as well as in supplying the growing plant with nutrients. A fertile soil may have 65% of these negative charges (cation exchange sites) occupied by Ca^{2+}, 10% by Mg^{2+}, 5% by K^+ and 20% by H^+. However, this is not a static situation; cations in the soil solution can replace bound cations and this replacement is called *cation exchange*. Cation exchange *in soils* is governed by two rules:

(a) The 'replacing power' of cations is as follows:

$$H^+ > *Al^{3+} > Ca^{2+} > Mg^{2+} > K^+ > Na^+$$

For example, *at equal concentrations*, K^+ ions will replace Ca^{2+} ions on the exchange sites more completely than will Na^+ ions, and so on. Overall, $H^+(H_3O^+)$ and $*Al^{3+}$ ions are the most efficient at dislodging cations. This order of cations is also the order of strength of binding to the exchange sites, i.e. H^+ will be more firmly bound than $*Al^{3+}$ etc.

(b) Law of Mass Action (section 7.6). The 'replacing power' of an ion depends also upon its *concentration* in the soil solution. Thus Na^+ at a high concentration can replace Ca^{2+} more completely than K^+ at a lower concentration.

These rules may be illustrated by considering what will happen to our fertile soil (see above) under cropping. Uptake of calcium by the roots of the crop plants will tend to lower the soil solution concentration of Ca^{2+} such that, by rules (a) and (b), H_3O^+ ions in the soil solution will begin to replace Ca^{2+} ions on the exchange sites (III) thereby releasing Ca^{2+} ions into the soil solution where they can, in turn, be lost from the soil by plant uptake or by leaching. This process can therefore continue until a large proportion of the exchange

---

*$*Al^{3+}$ ions take part in ion exchange only at low pH (section 5.3).

III

sites are occupied by H⁺ ions. At first sight it would seem that, according to rule (a), these sites should now be permanently occupied by H⁺ ions (the most firmly bound ions), giving a permanently acidic and infertile soil. However, we must not ignore rule (b). When we apply lime ($CaCO_3$ or $Ca(OH)_2$) to the soil we increase greatly the soil solution concentration of Ca^{2+} ions which, by rule (b), can now replace H⁺ ions on the exchange sites. Similar phenomena occur when potassium and ammonium fertilizers are applied to soils.

The Deionization of Water

The purification of water by ion exchange resins proceeds in two stages. First, the water is passed through a column containing a polymer (resin) carrying cation exchange sites saturated with H⁺ ions. These cation exchange sites have a higher affinity for metallic cations than for hydrogen (compare with soil, rule (a)), and, therefore, reactions like:

$$\boxed{\text{Resin}} - H + Na^+ + H_2O \rightarrow \boxed{\text{Resin}} - Na + H_3O^+$$

remove all the cationic impurities and release hydronium ions. The water is now passed through a second column containing a resin carrying anion exchange sites saturated with hydroxide ions. Reactions like:

$$\boxed{\text{Resin}} - OH + Cl^- \rightarrow \boxed{\text{Resin}} - Cl + OH^-$$

remove anions from solution and release hydroxide ions.

All anion and cation impurities have now been removed and the following neutralization reaction occurs:

$$H_3O^+ + OH^- \rightarrow 2H_2O.$$

Once all the exchange sites on the resin have become saturated with impurities, the columns can be regenerated using concentrated acid and alkaline solutions (according to rule (b)).

In general, the behaviour of an ion-exchanging material (soil, resin, etc.) towards a solution containing ionic solutes can be described by a replacement/affinity/binding series, as in rule (a), which is characteristic of the ion-exchanging material, modified by the Law of Mass Action.

EXERCISES

(1) What would be the consequences for the properties of water and for plants and animals if the Earth moved nearer to, or further from, the Sun?

(2) Under what temperature conditions could ammonia replace water in living organisms?

FURTHER READING

On the properties of water and their unique importance for life on Earth:

1. DAVIS K.S. & DAY J.A. (1961) *Water*. Heinemann, London.
2. FITTER A.H. & HAY R.K.M. (1981) *Environmental Physiology of Plants*, Ch. 4. (In preparation). Academic Press, London.
3. HENDERSON L.J. (1913) *The Fitness of the Environment*. Macmillan, London. (Reprinted by Beacon Press, Boston, 1958.) (An outstanding, historic, review of the ecological role of water.)
4. FOGG G.E. (Ed.) (1965) *The State and Movement of Water in Living Organisms.* Symposia of the Society for Experimental Biology 19. Cambridge University Press.

Reviews of the optical properties of water can be found in:

5. BAINBRIDGE R., EVANS G.C. & RACKHAM O. (1966) *Light as an Ecological Factor*, 1. Symposium of the British Ecological Society 6. Blackwell Scientific Publications, Oxford.
6. EVANS G.C., BAINBRIDGE R. & RACKHAM O. (1975) *Light as an Ecological Factor*, II. Symposium of the British Ecological Society 16. Blackwell Scientific Publications, Oxford.

For a rigorous study of water purification:

7. CAMP T.R. & MESERVE R.L. (1974) *Water and its Impurities*. Dowden, Hutchinson and Ross, Inc.

The fundamentals of ion exchange are discussed by:

8. PAULING L. (1970) *General Chemistry*, 3rd Edition, Ch. 12. W.H. Freeman, San Francisco.

For soils, it has been possible to combine ion exchange rules (a) and (b) into a single law—the Ratio Law. This law and subsequent developments (e.g. the Gapon Equation) are discussed in:

9. WHITE R.E. (1979) *Introduction to the Principles and Practice of Soil Science*, Ch. 7. Blackwell Scientific Publications, Oxford.
10. RUSSELL E.W. (1973) *Soil Conditions and Plant Growth*, 10th Edition, Ch. 7. Longman, London.

CHAPTER 10
SURFACE AND COLLOID CHEMISTRY

Several crops (e.g. cotton) must be sprayed very frequently with insecticide to prevent loss of yield and deterioration in the quality of the product. However, if we examine a cotton plant during rain, it is clear that raindrops do not adhere to the surfaces of the leaves but roll off almost immediately. How can it be, then, that the droplets of an insecticide spray do adhere to the leaf and spread out to give an even distribution of the dissolved chemical over its surface? In this chapter we shall try to answer this question, and also to explain some features of the surfaces of soil particles, by studying the physical chemistry of liquid/solid interfaces.

10.1 SURFACE TENSION

In the bulk of any liquid, each molecule is attracted equally in all directions to the surrounding molecules. However, at the surface, the outermost molecules tend to be pulled inwards (into the bulk of the liquid) because there is no molecular attraction from outside the liquid surface—

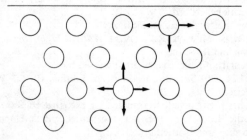

These inwardly directed forces cause small volumes of the liquid to assume the shape with the lowest surface area: volume ratio. This shape is a sphere—

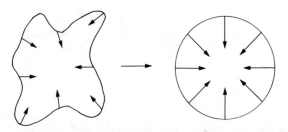

Once a volume of liquid has taken up a spherical shape, these inwardly directed forces resist any expansion in surface area. If we do wish to increase the surface area (e.g. to spread the droplet out as a thin layer covering a leaf), then work must be done against these forces, and the stronger the forces are, the greater the amount of work required to give unit increase in surface area.

The Surface Tension of a liquid is an index of the amount of work required to give unit increase in the surface area of a volume of liquid (measured under standard conditions). As shown in Table 10.1, liquids with unusually strong intermolecular forces (e.g. water) or interatomic forces (mercury) have high surface tensions.

TABLE 10.1. The surface tension of selected liquids in contact with air at 20°C.

	Surface tension $mN\ m^{-1}$
Water	72·75
Benzene	28·83
Acetic acid	27·60
Carbon tetrachloride	26·80
Acetone	23·70
Ethanol	22·30
Mercury	485·00

10.2 THE WETTING OF SURFACES—DETERGENTS

Up to this point, we have considered liquid droplets in air, but if we now place a water droplet on a hydrophilic surface (e.g. glass or cotton cloth) which carries hydroxyl groups—

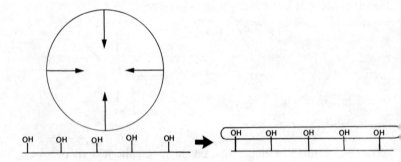

then hydrogen bonds can form between the outermost molecules in the water droplet and the hydroxyl groups on the hydrophilic surface. This 'neutralizes' the inwardly directed forces, reduces the surface tension of the water and enables the droplet to spread thinly over the surface. Thus the water in the droplet can *wet*, and adhere to, the surface.

On the other hand, if we place the water droplet on a hydrophobic surface with which it cannot form bonds, then the high surface tension of water results in *no wetting* of the surface. There is a minimum of contact between liquid and surface, and the droplet will roll off.

The successful wetting of hydrophobic surfaces is an important problem in the application of pesticides. In spraying a crop with insecticide or fungicide, it is necessary to cover the leaves uniformly with spray so that when the water evaporates, the active ingredient is uniformly distributed, giving reasonably complete protection. However, leaf surfaces are hydrophobic, due to a thin coating of wax over the epidermis, and, therefore, aqueous solutions and suspensions roll off without wetting the surface. Similarly, the hair and hides of livestock tend to be greasy and hydrophobic, repelling aqueous dip solutions.

It is not feasible to solve this problem by modifying the hydrophobic surfaces of crop plants and livestock animals, because they are important in protecting the epidermis against damage, desiccation or heat loss. Instead it is necessary to alter the surfaces of spray droplets so that they adhere to hydrophobic surfaces; this is achieved using detergents.

Detergents are amphipathic organic molecules containing:
(a) a hydrophilic functional group, attached to
(b) a long hydrophobic hydrocarbon chain (normally longer than 8 carbon atoms).

For example *soaps*, the sodium salts of long-chain carboxylic acids (section 13.6), are the most familiar detergent substances, e.g. sodium stearate:

$$CH_3-CH_2-CH_2-CH_2-CH_2-CH_2-CH_2-CH_2-CH_2-CH_2---$$
$$-CH_2-CH_2-CH_2-CH_2-CH_2-CH_2-CH_2-CO_2^-\ Na^+$$

or more concisely, $CH_3(CH_2)_{16}-CO_2^-\ Na^+$

and in diagram form—

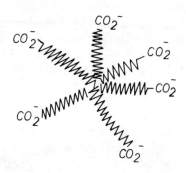

Synthetic Detergents are very similar to soaps but normally contain a sulphate ($-O-SO_3^-$), sulphonate ($-C_6H_4SO_3^-$) or, phosphate group instead of a carboxylate.

Although their molecules are predominantly hydrophobic, soaps and synthetic detergents dissolve in water due to the clustering of their molecules into *micelles*—

By this arrangement, all the hydrophobic chains are 'hidden' from the solvent, whereas all the hydrophilic groups are in contact with, and can form bonds with, water molecules. Thus, a detergent micelle can be thought of as a giant ion carrying many negative charges; in addition to these hydrophilic interactions between micelle and water, the mutual repulsion of the negatively charged micelles also favours the solubility of detergents (see section 10.5).

When a crop spray droplet containing a detergent (or 'spreader') comes into contact with a leaf, these micelles break up because the hydrophobic 'tails' of the detergent molecules are attracted to the hydrophobic surface of the leaf. However, because the hydrophilic ends of the detergent molecules remain firmly 'anchored' in the water of the droplet, the detergent molecules form a 'bridge' between the aqueous solution and the leaf surface, thereby overcoming the high surface tension of water—

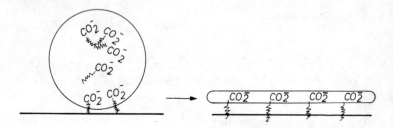

Consequently, the spray droplets adhere to, and cover, the leaf surface.

A more familiar example of detergent action is the removal of dirt from the hydrophobic surfaces of skin using soap and water. Water alone is not effective in cleaning the skin because of repulsion by the hydrophobic surface of skin and by hydrophobic dirt. However, the tails of soap molecules are attracted to greasy materials, whereas the hydrophilic ends remain in the water. When we rub our hands in soapy water, particles of dirt are detached from the skin at the centre of soluble micelles—

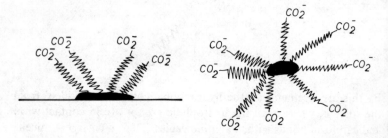

Soaps cannot be used in 'hard' water (section 9.4) because calcium, magnesium and iron cations cause the precipitation and inactivation of the soap anions, e.g.

$$Ca^{2+} + 2CH_3(CH_2)_{16}CO_2^- \rightarrow Ca(CH_3(CH_2)_{16}CO_2)_2\downarrow.$$
$$\text{calcium stearate}$$

Instead, a synthetic detergent which does not precipitate may be used. Detergents have been devised for a wide range of applications including the dispersion of oil pollution in the sea.

10.3 INTERACTIONS BETWEEN SOIL SURFACES AND WATER

Soils consist of mineral particles (mostly hydrophilic) of varying diameter (sand 2–0·06 mm, silt 0·06–0·002 mm and clay < 0·002 mm, Soil Survey of England and Wales classification), bound together into aggregates by organic matter and clay particles. Within and between these aggregates, there is a network of interconnected pores of diameter ranging from *a few cm* (drying cracks, earthworm or termite channels), through *a few mm* (pores between aggregates), down to *a few μm or tenths of a μm* (finest pores within aggregates). Thus the internal surface area: volume ratio of a well-aggregated soil is enormous.

When a dry soil is first wetted, water is attracted to the hydrophilic surfaces of soil particles and tends to be spread thinly over a very large surface area. However, as we have seen in section 10.1, water tends to change its shape so as to *minimize* its surface area in contact with air. In soils, this reduction in the air/water interface is achieved by the *filling* of pores with water; once a soil pore has been filled, work must be done against the surface tension of water to withdraw water from the pore, because withdrawal involves an increase in the water surface in contact with air.

For example, after the drainage of a temperate soil is complete, all soil pores narrower than 60 μm (10 μm in tropics) will be filled with water, bound by the hydrophilic pore walls and by the surface tension of water. As pores become narrower, these retaining forces (normally called the *Matric* forces) increase as shown by the expression:

$$\frac{\text{Suction required to withdraw water}}{\text{from pore (bars)}} = \frac{3}{d}$$

where d is the pore diameter in μm.

Another feature of soil/water relationships originating from surface properties is the *capillary movement* of water. This phenomenon can be explained using the simpler glass/water system. For example, if we insert the end of a wide bore (several mm in diameter) glass tube into a beaker of water, water molecules are attracted to, and pulled up, the hydrophilic internal surface of the tube, giving the familiar concave, rather than flat, meniscus. However, there is no upward movement of the bulk of the water inside the tube—

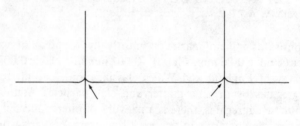

If we reduce the internal diameter of the glass tube, then the *proportion* of water molecules in contact with the internal surface of the tube increases whereas the mass of the column of water in the tube decreases. Consequently, at a certain diameter, the upward-directed forces between glass and water become large enough to raise the water column—

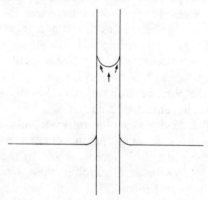

For example, an internal diameter of 1 mm gives a capillary rise of 30 mm above the water level in the beaker compared with 1·5 m with a diameter of 0·02 mm. Such large rises are possible only because the strong intermolecular forces in water allow transmission of the force from the molecules in contact with the glass to the bulk of the liquid.

Capillary rise, or capillarity, is responsible for much of the vertical and lateral movement of water in soils. The role of capillary action in bringing water to the soil surface from a deep water table has been the subject of considerable controversy, but it can be demonstrated in the field that capillary rise can subirrigate crops adequately from a water table at a depth of at least 1 metre.

10.4 THE COLLOIDAL STATE

If we take a cube of material of side 1 m and cut it into 8 identical cubes, each of side $\frac{1}{2}$ m, then the total mass and volume of material remains the same but the surface area has increased from 6 m² to 12 m². If each of these smaller cubes is similarly divided, we obtain 64 cubes of total surface area 24 m². Thus the surface area: volume ratio is doubled each time the side is halved (Table 10.2). Consequently, if a substance is divided up very finely, its surface area per unit volume (or mass) becomes very large and the properties of its surface come to dominate its chemical behaviour. For example many powders, such as finely ground flour, can burn spontaneously in air although the bulk material requires activation energy to be supplied before combustion can take place. Similarly, the behaviour of a *colloid* tends to be determined more by its surface area than by the chemical properties of the material in bulk.

TABLE 10.2. The effect of subdivision of a solid cube on surface area.

Cube side	No. of cubes	Total volume (m³)	Total surface area (m²)	Surface area / volume
1 m	1	1	6	6
$\frac{1}{2}$	8	1	12	12
$\frac{1}{4}$	64	1	24	24
$\frac{1}{8}$	512	1	48	48
and so on.				

Matter exists in the *Colloidal State* when it is divided up into small particles of diameter 10^{-6} to 10^{-9} m; thus colloids are defined by the *size* of their particles, which may consist of very large molecules (macromolecules) or clusters of molecules or atoms.

Colloidal Systems consist of matter in the colloidal state (the *Dispersed Phase*) spread uniformly throughout a *Dispersion Medium*. In theory, both dispersed phase and dispersion medium can exist in each of the three states of matter, but the more familiar combinations are shown in Table 10.3. In the following section, we shall concentrate on sols and emulsions which are the most important kinds of colloidal system in environmental chemistry.

TABLE 10.3. Some familiar colloidal systems.

Dispersed phase	Dispersion medium	Colloidal system	Example
Liquid	Gas	Liquid aerosol	Fog
Solid	Gas	Solid aerosol	Smoke, dust
Gas	Liquid	Foam	Soap lather, fire extinguisher foam
Liquid	Liquid	Emulsion	Milk
Solid	Liquid	Sol, paste	Toothpaste

10.5 COLLOIDAL SYSTEMS

It is useful to recognize three distinct classes of sols (and emulsions).

(a) *True solutions of macromolecules*. The most familiar examples of this type of sol are solutions of hydrophilic macromolecules (proteins, carbohydrates) in water. In particular, the cytoplasm is a complex solution of many different macromolecules, predominantly proteins. Where the dispersion medium is other than water (e.g. rubber in benzene), the dispersed phase is called *Lyophilic*, solvent loving.

As well as possessing many of the normal properties of solutions, solutions of macromolecules have other characteristics; in particular:

High Viscosity. This accounts for the name colloid which means glue-like.

Dialysis. Unlike smaller solutes, macromolecules in solution do not pass through the pores of dialysis membranes. Consequently, dialysis is a useful technique in biochemistry for separating macromolecules from other, usually inorganic, solutes.

(b) *Association Colloid Systems* are formed when a number of amphipathic molecules (containing lyophobic and lyophilic groups) cluster together to form *micelles* of colloidal dimensions. As shown in section 10.2, soaps and detergents form association colloid systems in water.

(c) *Colloidal Dispersions*. Here the dispersed phase consists of atoms or molecules clumped together into particles of colloidal size. These are not true solutions because the dispersed phase is normally not strongly lyophilic; however, the clumps are not forced together by the solvent to give a precipitate because they tend to accumulate (charged) ions from solution on their surfaces. As a result of these charges, the clumps repel one another strongly and remain in suspension.

The hydrated iron oxide/water sol is an easily prepared colloidal dispersion, often demonstrated in the laboratory. It can be prepared by boiling a ferric chloride solution to give hydrated iron oxide $Fe_2O_3.xH_2O$ which is insoluble in water. However, as the molecules of iron oxide begin to clump together, they absorb hydronium ions from solution, giving a positive surface charge—

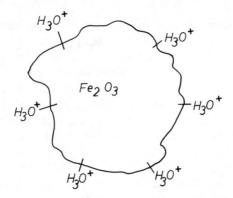

which repels other positively charged colloidal particles; the dispersion is, therefore, stabilized by the repulsion of neighbouring colloid particles. The addition of large anions like the ferricyanide ion $(Fe(CN)_6^{3-})$ causes the iron oxide to precipitate (or *flocculate*) by neutralizing the positive charges.

Silicate Clays

As we noted in section 4.4, silicate clays are similar in structure to sheet silicates but with aluminium hydroxide layers interposed between the silicate sheets. However, in contrast to the large flaky crystals of mica, silicate clays in soils exist predominantly in the colloidal state. In addition, these colloidal particles are negatively charged, due to the isomorphous replacement of Al for Si and Mg for Al (see section 4.4). Some important properties of three common clay minerals are given in Table 10.4.

Soil clay particles, therefore, have enormous surface areas upon which occur many of the important reactions of soil chemistry. In particular, these surfaces (together with the surfaces of humus

TABLE 10.4. Selected properties of silicate clay minerals in soils. (From *The Nature and Properties of Soils*, 8th Edition, by N.C. Brady. © 1974 by Macmillan Publishing Co. Inc.)

Clay mineral	Particle size	Surface area	Negative charge
Kaolinite	$0 \cdot 1 - 5 \cdot 0 \, \mu m$	$5–20 \, m^2 \, g^{-1}$	$3–15 \, mEq^* \, 100 \, g^{-1}$
Illite	$0 \cdot 1 - 2 \cdot 0$	100–120	15–40
Montmorillonite	$0 \cdot 01 - 1 \cdot 0$	700–800	80–100

* Using the ion-exchange definition of an equivalent (section 8.3).

particles, also colloids) are the sites of cation exchange (sections 8.2 and 9.5) and of deposition of the products of the weathering of soil minerals (e.g. iron oxides).

It is essential for the particles (sand, silt and clay) of agricultural soils to bind together into stable aggregates, since this promotes drainage and aeration and increases the amount of water available to plant roots. The primary process in the formation of aggregates is the flocculation of colloidal clay particles into *domains* which, with humic material, bind the sand and silt particles together.

As with micelles and colloidal dispersions, clay particles in water tend to remain dispersed because of mutual repulsion of their negatively charged surfaces. However, as we saw in section 9.5, in soil these negative charges are neutralized by cations. The idea of cation exchange *sites* is useful when discussing cation exchange but when considering the flocculation of clay, it is more helpful to consider the negative charge to be spread over the whole colloid particle, and shielded from neighbouring particles by a 'cloud' of cations. (The negative charge at the colloid surface and its accompanying cloud of cations are together normally called the *diffuse double layer*.) Cations differ in their ability to shield the negative charges on colloid particles according to the series:

$$Al^{3+} > Ca^{2+}, H^+ > Mg^{2+} > K^+ > Na^+$$

For example, Ca^{2+} ions are so successful in shielding the negative charges on adjacent clay particles that the particles cease to repel one another and flocculation can occur. On the other hand Na^+ ions are much less successful, causing the clay to remain dispersed or deflocculated.

Consequently, soils with a high Ca^{2+} ion content, maintained by liming, tend to have good physical properties due to the aggregation of soil particles. In contrast, in saline soils, the deflocculation of clay leads to poor aggregation and unfavourable physical conditions for crop growth.

EXERCISES

(1) Why does a column of mercury in a glass tube have a convex meniscus?

(2) Milk is a colloidal dispersion of hydrophobic lipid in water. How is the lipid flocculated to give butter?

FURTHER READING

On the physical chemistry of surfaces and colloids:
1. HOFFMAN K.B. (1963) *Chemistry for the Applied Sciences*, Ch. 10. Prentice-Hall, Inc, New Jersey.
2. SHAW D.J. (1970) *Introduction to Colloid and Surface Chemistry*, 2nd Edition. Butterworth, London.
3. VOLD M.J. & VOLD R.D. (1965) *Colloid Chemistry*, Van Nostrand Reinhold, Wokingham.
On the use of 'spreaders' in crop protection, consult:
4. MARTIN H. (1973) *The Scientific Principles of Crop Protection*, Ch. 3. Edward Arnold, London.
On the properties of soil colloids:
5. WHITE R.E. (1979) *Introduction to the Principles and Practice of Soil Science*. Chs 2, 3, 4 and 7. Blackwell Scientific Publications, Oxford.

PART 2
ORGANIC CHEMISTRY

CHAPTER 11

INTRODUCTION TO ORGANIC

CHEMISTRY

11.1 THE UNIQUENESS OF CARBON

Organic chemistry is the study of covalently bonded carbon compounds, although ionic bonds do occur in a few organic substances. One of the reasons for studying carbon compounds separately from those of the other 102 (or more) elements is simply that there are more compounds of carbon than of all the other elements put together. It has been calculated that there are 3 million known organic compounds compared with about 60 000 inorganic compounds. A more important reason is that all living organisms are made up principally of organic substances, and use a vast array of carbon compounds in their growth and physiology.

Carbon forms more compounds than any other element due to a unique combination of three atomic properties.

Property A

Carbon has a high valency, 4. To obtain a stable octet, each carbon atom can share a pair of electrons with four other atoms, giving four single bonds, for example, in methane, CH_4—

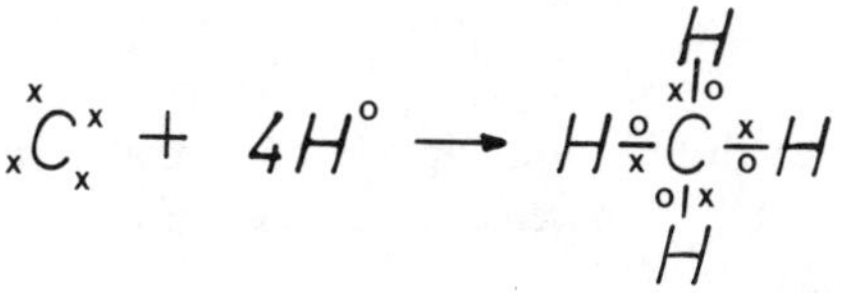

$$_x^x C_x^x \ + \ 4H° \ \longrightarrow \ H \overset{x|o}{\underset{o|x}{\circ\overset{}{x}}} C \overset{}{\underset{}{\circ}} H$$

(Note that although the bonds in singly bonded carbon atoms are directed towards the corners of a tetrahedron, it is normally more convenient in organic chemistry to draw them in one plane.) As we shall see in exercise 2 (page 130), if carbon forms bonds with a number

127

of different atoms, its high valency alone can result in a large number
of possible compounds.

Property B

Single covalent bonds linking carbon atoms are strong not only in the
pure element (see section 4.2) but also in compounds of carbon with
other elements, especially hydrogen. This permits the formation of
chains of carbon atoms, both straight and branched and of un-
limited length, as well as the formation of rings of carbon atoms, for
example—

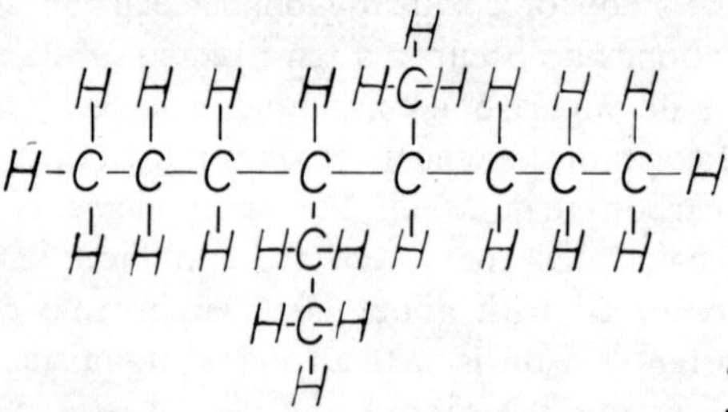

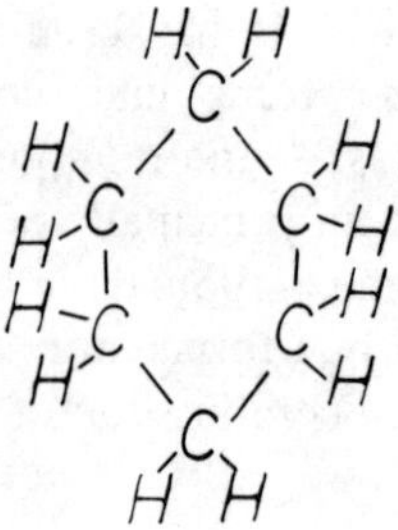

The atoms of several other elements (S, P, etc.) are linked in groups
by strong covalent bonds, in the element, but not when their atoms
are attached to the atoms of other elements. Because of this linking
of carbon atoms into chains and rings of different sizes and shapes,
an enormous number of compounds of carbon is possible.

Property C

Carbon to carbon double and triple bonds are also strong and stable,
resulting in a further increase in the variety of possible organic
compounds, for example—

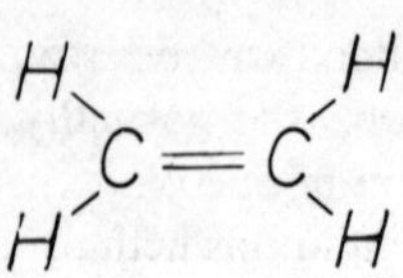

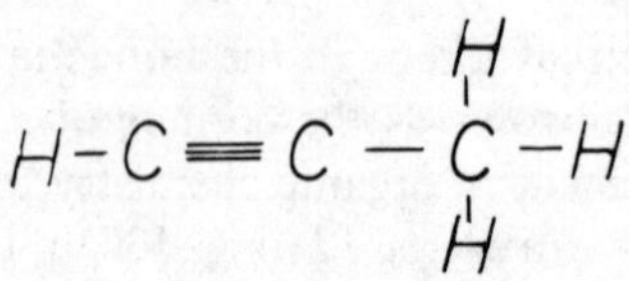

The atoms of several elements have at least one of these properties but carbon atoms are unique in possessing all three. As a result of A, B and C, we can predict the existence of an enormous number of even the simplest organic compounds—the *hydrocarbons*, which contain carbon and hydrogen atoms only.

11.2 HOMOLOGOUS SERIES AND FUNCTIONAL GROUPS

In spite of the large number of known organic compounds, it is not necessary to examine the properties and reactions of each compound individually. Instead, organic compounds can be classified into a few *homologous series* of closely related compounds with similar properties and reactions. For example, the series of compounds—

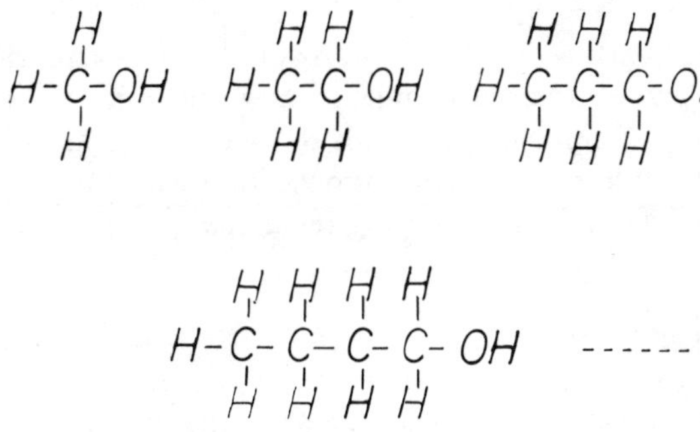

$$
\begin{array}{ccc}
\overset{H}{\underset{H}{H-C-OH}} & \overset{H}{\underset{H}{H-C-}}\overset{H}{\underset{H}{C-OH}} & \overset{H}{\underset{H}{H-C-}}\overset{H}{\underset{H}{C-}}\overset{H}{\underset{H}{C-OH}}
\end{array}
$$

$$
\overset{H}{\underset{H}{H-C-}}\overset{H}{\underset{H}{C-}}\overset{H}{\underset{H}{C-}}\overset{H}{\underset{H}{C-OH}} \quad ------
$$

is a homologous series of *alcohols*, each containing the hydroxyl (OH) *functional group* attached to a hydrocarbon chain of varying length. As we shall see in Chapter 12, hydrocarbon chains tend to be chemically inert and, therefore, the chemical properties of the alcohols depend mainly upon the hydroxyl functional group. Consequently, we can establish generalizations about the properties of alcohols by studying a few representative members of the series.

This simplifies the study of organic chemistry considerably, since, for example, a complex molecule containing a hydroxyl group will have at least some of the characteristic properties of the alcohols. Note, however, that although the length and branching of the hydrocarbon chain may have only a small effect upon the chemistry of a molecule, it will have a large effect upon the physical properties, especially boiling point and freezing point, and upon the water solubility. This will be considered in more detail in subsequent chapters.

In general, an organic compound consists of a hydrocarbon framework (chains and/or rings) with one or more functional groups attached. (No functional groups in the case of the hydrocarbon homologous series.) Some common functional groups are:

$$\text{hydroxyl} - \text{OH} \qquad \text{carboxylic acid} - \overset{\displaystyle O}{\overset{\displaystyle \|}{C}} - \text{OH}$$

$$\text{aldehyde} - \overset{\displaystyle O}{\overset{\displaystyle \|}{C}} - \text{H} \qquad \text{amine} \qquad - \text{NH}_2$$

$$\text{ketone} \quad - \overset{\displaystyle O}{\overset{\displaystyle \|}{C}} - \qquad \text{thiol} \qquad - \text{SH}.$$

Since the hydrocarbon framework normally has the same chemical properties whether attached to functional groups or not, we shall begin our study of organic compounds by establishing the properties and reactions of the four major groups of hydrocarbons before going on to study molecules carrying functional groups.

EXERCISES

(1) Compare the valency of carbon with other familiar elements in the periodic table. Does any element have the same, or a higher valency?

(2) By writing down chemical formulae, work out how many different compounds can be formed between a single carbon atom and atoms of the elements H, F, Cl, Br and I, e.g. CH_4, CH_3Cl, CH_2Cl_2, $CHCl_3$, CCl_4, $CHClFI$, etc.

FURTHER (BACKGROUND) READING FOR ORGANIC CHEMISTRY

1. HOFFMAN K.B. (1963) *Chemistry for the Applied Sciences*, Chs 14–25. Prentice-Hall, Inc, New Jersey.

2. HOLUM J.R. (1969) *Introduction to Organic and Biological Chemistry.* John Wiley & Son, New York.

3. ROSE S. (1966) *The Chemistry of Life*, Chs 1–3. Penguin Books, Middlesex.

4. ROSSOTTI H. (1975) *Introducing Chemistry*, Ch. 12. Penguin Books, Middlesex.

5. WHITE E.H. (1970) *Chemical Background for the Biological Sciences*, 2nd Edition, Chs 3–6. Prentice-Hall, Inc, New Jersey.

CHAPTER 12
THE HYDROCARBONS

The hydrocarbons, which contain C and H atoms only, may be sub-divided into four homologous series:

(1) The Alkanes (or Paraffins)
(2) The Alkenes (or Olefins)   together called the Aliphatic Hydrocarbons
(3) The Alkynes (or Acetylenes)
(4) The Arenes (or Aromatic Hydrocarbons)

according to the nature of their carbon to carbon bonds, as described in the following sections.

12.1 THE ALKANES

The alkanes are the simplest organic compounds since they are *saturated* (contain no double or triple bonds). Their general formula is C_nH_{2n+2}, for example—

Methane, CH_4

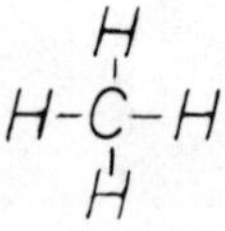

$$H-\overset{\overset{\displaystyle H}{|}}{\underset{\underset{\displaystyle H}{|}}{C}}-H$$

n-pentane, C_5H_{12}

$$H-\overset{\overset{\displaystyle H}{|}}{\underset{\underset{\displaystyle H}{|}}{C}}-\overset{\overset{\displaystyle H}{|}}{\underset{\underset{\displaystyle H}{|}}{C}}-\overset{\overset{\displaystyle H}{|}}{\underset{\underset{\displaystyle H}{|}}{C}}-\overset{\overset{\displaystyle H}{|}}{\underset{\underset{\displaystyle H}{|}}{C}}-\overset{\overset{\displaystyle H}{|}}{\underset{\underset{\displaystyle H}{|}}{C}}-H$$

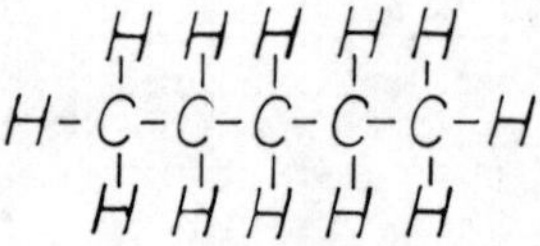

Up to this point, we have drawn each organic molecule in full, as for n-pentane above. However, it is much more convenient to use the following condensed forms:

$$CH_3—CH_2—CH_2—CH_2—CH_3 \text{ or } CH_3CH_2CH_2CH_2CH_3.$$

Some other members of the alkane homologous series are given in Table 12.1.

The Naming of Alkanes: Structural Isomers

The first four members of the alkane series, methane (1C), ethane (2C), propane (3C) and butane (4C) have special names. The remaining members are named by adding the suffix –ane to the Greek word for the number of carbon atoms in the hydrocarbon chain. Thus we have:

> pent -ane (5C)
> hex -ane (6C)
> hept -ane (7C)
> oct -ane (8C)
> non -ane (9C)
> dec -ane (10C)
> pentadec-ane (15C) etc.

When an alkane chain is attached to a functional group, we replace the -ane suffix with an -yl suffix,

> e.g. C_2H_5—Cl ethyl chloride
> C_6H_{13}—Cl hexyl chloride.

Groups like $C_2H_5^-$ and $C_6H_{13}^-$ are called *alkyl* groups. This naming system is quite adequate for straight-chain alkanes, but as we saw in section 11.1, branched chains can also exist. In the cases of methane, ethane and propane, we can arrange the carbon atoms only in straight chains:

$$CH_4 \quad CH_3—CH_3 \quad CH_3—CH_2—CH_3.$$

However, when we come to butane, C_4H_{10}, the carbon atoms can be arranged in two ways:

> (a) $CH_3—CH_2—CH_2—CH_3$
>
> $$CH_3$$
> $$|$$
> (b) $CH_3—CH—CH_3.$

(a) and (b) are not identical molecules. They have the same formula C_4H_{10} but different structures and are called *structural isomers* of butane. In general, structural isomers tend to have similar chemical properties but different physical properties.

We may call (a) n-butane and (b) iso-butane where n (or normal) indicates a straight chain and iso indicates the presence of—

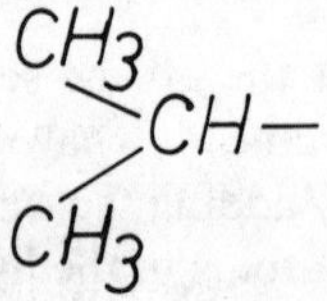

$$CH_3 \diagdown CH- \diagup CH_3$$

at the end of the chain. However, there exists a more systematic method of naming alkanes (and all other organic compounds) which may be explained using the following complex alkane as an example—

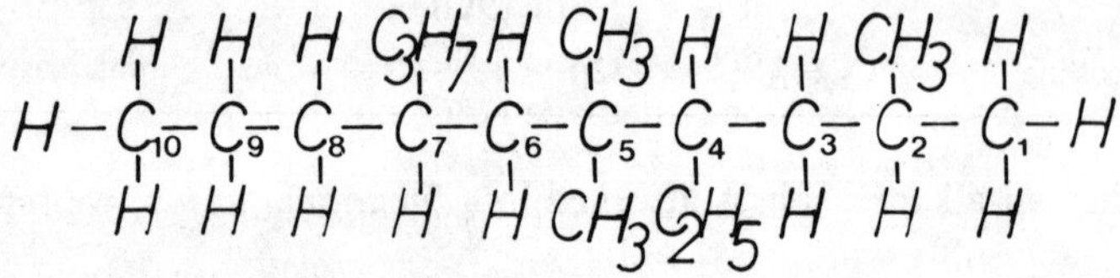

$$H-\underset{10}{C}-\underset{9}{C}-\underset{8}{C}-\underset{7}{C}-\underset{6}{C}-\underset{5}{C}-\underset{4}{C}-\underset{3}{C}-\underset{2}{C}-\underset{1}{C}-H$$

Step 1: Select the longest hydrocarbon chain in the molecule and name the molecule accordingly. Therefore, the above alkane is a *decane* (10C chain).

Step 2: Number the C atoms in this chain from 1 upwards.

Step 3: List the groups attached to the chain in order, indicating to which C atom they are attached, in the following way:

2-methyl-4-ethyl-5, 5-dimethyl-7-propyl-decane

Note that it would be incorrect to use the name:

4-propyl-6, 6-dimethyl-7-ethyl-9-methyl-decane

(numbering from the other end) since the position of branches (and functional groups) must be indicated by the lower set of numbers (i.e. 2, 4, 5 and 7 rather than 4, 6, 7 and 9).

Using this systematic method, we should call the (a) isomer of butane simply *butane* and the (b) isomer, *2-methyl-propane*.

Properties and Reactions of Alkanes

1. PHYSICAL PROPERTIES

All alkanes tend to have rather similar chemical properties, but their physical properties are more variable. In particular, as shown in

TABLE 12.1. Properties of some straight-chain alkanes.

Name	Formula	Molecular weight	m.p. °C	b.p. °C	Name of alkyl group
Methane	CH_4	16	-183	-164	methyl
Ethane	C_2H_6	30	-183	-89	ethyl
Propane	C_3H_8	44	-190	-42	n-propyl
Butane	C_4H_{10}	58	-138	-0.5	n-butyl
Pentane	C_5H_{12}	72	-130	36	n-pentyl
Hexane	C_6H_{14}	86	-95	69	n-hexyl
Heptane	C_7H_{16}	100	-91	98	n-heptyl
Octane	C_8H_{18}	114	-57	126	n-octyl
Decane	$C_{10}H_{22}$	142	-30	174	n-decyl
Octadecane	$C_{18}H_{38}$	254	28	316	n-octadecyl
Eicosane	$C_{20}H_{42}$	282	37	343	n-eicosyl

Table 12.1, their melting and boiling points depend upon the number of carbon atoms in the formula, the values rising as the molecules become larger and heavier. As a result, the first four alkanes (methane to butane) are gases, the next eleven (pentane to pentadecane) are liquids and the remaining members are solids at room temperature.

2. HYDROPHOBIC NATURE

Alkane (and other hydrocarbon) molecules are non-polar and highly hydrophobic. In the presence of water, they attract one another because they are strongly repelled by polar water molecules. This mutual attraction of hydrocarbon molecules, which is sometimes called hydrophobic bonding accounts for the very low water solubility of alkanes.

The hydrophobic nature of alkane molecules and alkyl groups has many important consequences throughout biology and agriculture. For example, hydrophobic bonding is involved in the behaviour of detergents (section 10.2) and in determining the tertiary structures of proteins (section 14.2). At another level, the disastrous consequences for seabirds of oil spills at sea are a result of the attraction between hydrocarbons and the hydrophobic lipids on their feathers.

3. COMBUSTION

Because the complete combustion (burning) of an alkane in the presence of adequate oxygen is a highly exergonic reaction, alkanes are useful fuels for heating, lighting and fuelling engines. For example the combustion of methane,

$$CH_4 + 2O_2 \rightarrow CO_2 + 2H_2O,$$

releases $8\cdot9 \times 10^5$ J mole^{-1} of methane (213 Kcal mole^{-1}).
Other important fuels include:

Natural or Bottled Gas—(for household, laboratory and industrial heating) contains C1 to C4 alkanes;

Gasoline or Petrol —(for internal combustion engines) contains C6 to C9 alkanes;

Kerosine or Paraffin —(for jet engines and heating) contains C10 to C16 alkanes.

Note that if insufficient oxygen is supplied during combustion, products other than CO_2 and H_2O are formed (e.g. carbon, soot) and less energy is released.

Petrol must not contain too high a proportion of straight-chain alkanes because these tend to cause explosions ('knock') in the cylinders of the engine, giving uneven power output. The 'octane number' of a given petrol supply is a measure of how highly branched the fuel is and is obtained by comparing the performance of the fuel with the highly branched octane 2,2,4-trimethyl pentane; the higher the octane number, the more even the power output. Most petrol supplies also contain an 'anti-knock' additive, such as tetraethyl lead, $Pb(C_2H_5)_4$ which is the source of significant amounts of lead pollution (air, soil and water) in urban areas and near motorways.

4. CHEMICAL UNREACTIVITY

In living tissues, alkanes and alkyl groups are highly unreactive, although under industrial conditions they can undergo a variety of reactions. Consequently, most biochemical reactions involve functional groups rather than the rather inert hydrocarbon framework.

The Occurrence of Alkanes in Nature

The most important source of alkanes is petroleum or crude oil from oil wells in various parts of the world. Petroleum is a complex mixture of many organic compounds which can be divided by distillation into a number of useful fractions, each fraction containing several similar compounds. Elsewhere in nature, alkanes are uncommon although they do occur as waxes in the hydrophobic cuticles of leaves.

Halogenated Alkanes

The replacement of hydrogen atoms in simple alkanes by halogen atoms (F, Cl, Br, I) gives rise to a number of useful compounds including:

Chloroform (Trichloromethane)	$CHCl_3$	—Anaesthetic
Carbon tetrachloride (Tetra-chloromethane)	CCl_4	—Dry cleaning solvent
Methyl bromide	CH_3Br	—Used in soil sterilization
Iodoform (Tri-iodomethane)	CHI_3	—Antiseptic and local anaesthetic
Ethyl chloride	C_2H_5Cl	—Local anaesthetic
γ-BHC (Hexachloro-cyclohexane)	$C_6H_6Cl_6$	—Organochlorine insecticide

Halogenated alkanes, like alkanes, are particularly resistant to chemical attack. Consequently organochlorine pesticides like γ-BHC tend to accumulate in the environment and in fat stores in living tissues (section 16.1).

12.2 THE ALKENES

The alkenes are straight- and branched-chain hydrocarbon compounds containing one or more carbon to carbon double bond. They are, therefore, *unsaturated* and their general formula (1 double bond in the molecule) is C_nH_{2n}, e.g.

Ethene (Ethylene)	$CH_2{=}CH_2$
Propene (Propylene)	$CH_2{=}CH{-}CH_3$.

TABLE 12.2. Properties of some straight-chain alkenes.

Name	Formula	Molecular weight	m.p. °C	b.p. °C
Ethene	C_2H_4	28	−169	−104
Propene	C_3H_6	42	−185	−47
1-butene	C_4H_8	56	−185	−6
1-pentene	C_5H_{10}	70	−138	30
1-hexene	C_6H_{12}	84	−140	63
1-decene	$C_{10}H_{20}$	140	−66	171
1-octadecene	$C_{18}H_{36}$	252	18	179
1-eicosene	$C_{20}H_{40}$	280	29	341

Other members of the C_nH_{2n} series are given in Table 12.2. As with the alkanes, the melting points and boiling points of alkenes depend (somewhat irregularly) upon the length of the carbon chain. Note that the insertion of a double bond into a molecule gives a lowering of m.p. and b.p. (for example, compare hexane with 1–hexene in Tables 12.1 and 12.2) such that long-chain alkenes tend to be liquids at room temperature whereas the corresponding alkanes are solids (compare octadecane with 1-octadecene). This effect is important in determining the states of edible lipids (see below).

The Naming of Alkenes

As we can see in Table 12.2, alkene compounds are named by sub-stituting the suffix -ene for -ane in the name of the corresponding alkane, e.g.

$$CH_3—CH_3 \qquad CH_2{=}CH_2$$
ethane $\qquad\qquad$ ethene
$$CH_3—CH_2—CH_3 \quad CH_2{=}CH—CH_3$$
propane $\qquad\qquad$ propene.

However, more than one isomer exist for higher members of the series since the double bond may be in different positions, e.g.

$$CH_3—CH_2—CH_2—CH_3 \quad (a)\; CH_2{=}CH—CH_2—CH_3$$
n-butane $\qquad\qquad\qquad (b)\; CH_3—CH{=}CH—CH_3$
$\qquad\qquad\qquad\qquad\qquad (c)\; CH_3—CH_2—CH{=}CH_2$
$\qquad\qquad\qquad\qquad\qquad$ butenes.

Here, (a) is identical to (c) but (b) is a different isomer. (a) is, there-fore, called *1-butene* and (b) *2-butene*—the number indicating the position of the double bond. Groups attached to a main chain can be indicated as in the naming of alkanes. Thus,

$$CH_3$$
$$|$$
$$CH_2{=}CH—CH_2—CH—CH_2—CH_3 \text{ is called 4-methyl-1-hexene.}$$

Properties and Reactions of Alkenes

Alkenes are very similar to alkanes in their hydrophobic properties and high heats of combustion; however, the presence of double bonds makes alkenes much more reactive.

1. Saturation reactions

The most characteristic reactions of alkenes are *saturation* reactions which involve the *addition* of functional groups to double-bonded carbon atoms, thus removing the double bond. In biochemistry and nutrition, there are two important types of saturation, Hydrogenation (addition of hydrogen) and Halogenation (addition of halogens, especially bromine and iodine).

(a) Hydrogenation, e.g.

$$CH_3—CH{=}CH—CH_3 + H_2 \rightarrow CH_3—CH_2—CH_2—CH_3.$$

Hydrogenation reactions of alkenes have important applications in the margarine industry. The raw materials for the manufacture of margarine are oils (*liquid* lipids, e.g. groundnut, sunflower or cottonseed oil) which are made up of large molecules containing long, unsaturated hydrocarbon chains. These oils can be 'hardened' (i.e. converted into more useful solid fats) by the saturation of the hydrocarbon chains with hydrogen, normally with the help of a finely divided platinum catalyst.

Unsaturated fatty acids (see section 13.6) are an essential part of the diet of both humans and livestock, e.g.

$$CH_3—CH_2—CH_2—CH_2—CH_2—CH{=}CH—CH_2—CH{=} ----$$
$$---- CH—CH_2—CH_2—CH_2—CH_2—CH_2—CH_2—CH_2—CO_2H$$

Linoleic acid.

(b) Halogenation, e.g.

$$CH_3—CH{=}CH—CH_3 + Br_2 \rightarrow CH_3—\overset{\displaystyle Br}{\overset{|}{C}}H—\overset{\displaystyle Br}{\overset{|}{C}}H—CH_3.$$

This reaction is the basis of the standard qualitative test for the presence of multiple bonds (unsaturation) in organic compounds. Bromine dissolved in carbon tetrachloride is added dropwise to the substance under test. An unsaturated compound will react with the added bromine resulting in the disappearance of the characteristic red-brown colour of bromine.

It is essential for the food industry to have a quantitative estimate of the amount of unsaturation in lipid molecules; this is normally expressed as the *Iodine Number*—the mass of iodine absorbed (by addition reactions) per 100 g of sample of fat or oil.

2. OXIDATION REACTIONS

Under certain conditions, carbon to carbon double bonds in large molecules can react with atmospheric oxygen to give a complex mixture of products. In unsaturated oils, the products have unpleasant tastes and smells. A fatty material oxidized in this way is said to be rancid and may be unfit for human consumption.

3. POLYMERIZATION REACTIONS

Two important plastics, used widely in everyday life and agriculture, are manufactured from simple alkene molecules. *Polythene* (or polyethylene) is made by the linking of many ethene molecules into long, branched chains (polymerization), i.e.

$$CH_2{=}CH_2 \quad CH_2{=}CH_2 \quad CH_2{=}CH_2 \text{ (monomers)}$$
$$-CH_2-CH_2-CH_2-CH_2-CH_2-CH_2-, \text{ etc. (polymer).}$$

Polythene is the familar plastic used for pipes, buckets, plastic bags, etc.

Polypropylene is made in a similar fashion but starting with propene as the monomer. Polypropylene is stronger than polythene and has a higher melting point. Woven polypropylene is a strong and durable substitute for hessian (jute).

Some important natural compounds containing alkene chains

(1) *Ethene* is a powerful plant growth regulator, sometimes used commercially to promote ripening in fruit.

(2) *Fatty acids*, see above, and in section 13.6.

(3) *Many highly coloured compounds* occurring in fruit and vegetables are alkenes, e.g. Lycopene, the red colouring matter in tomatoes—

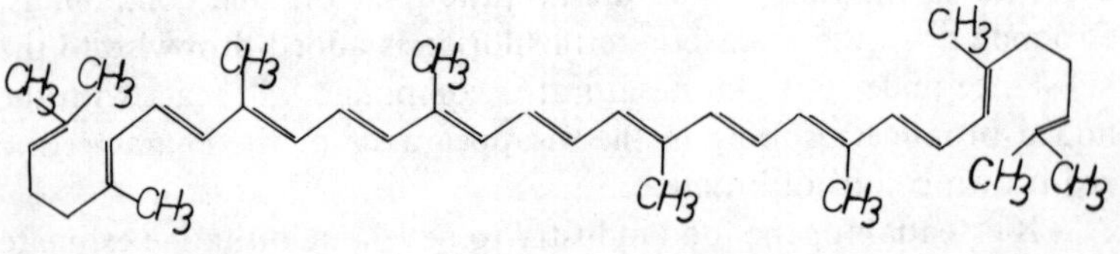

(where the C and H atoms on the main chain are omitted, for clarity).

(4) *Vitamin A1 or Retinol* (deficiency in mammals results in retarded growth and night blindness—

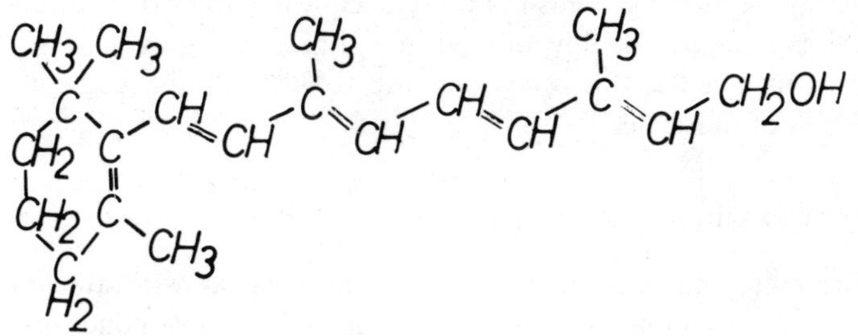

Vitamin A1

12.3 THE ALKYNES

The alkynes are straight- and branched-chain hydrocarbon compounds containing one or more triple bonds. Their general formula (1 triple bond in the molecule) is C_nH_{2n-2}, e.g.

$$\text{Ethyne (Acetylene)} \quad CH \equiv CH$$
$$\text{Propyne} \quad CH \equiv C-CH_3.$$

The Naming of Alkynes

Alkyne compounds are named by substituting the suffix -yne for -ane in the name of the corresponding alkane, e.g.

$$CH_3-CH_3 \text{ ethane} \quad CH \equiv CH \text{ ethyne.}$$

The position of the triple bond in higher members of the series is indicated in the same way as in alkenes, i.e.

$$CH_3-CH_2-CH_2-CH_3 \quad CH \equiv C-CH_2-CH_3$$
butane   1-butyne
$$CH_3-C \equiv C-CH_3$$
2-butyne.

Properties and Reactions of Alkynes

The presence of triple bonds in alkynes makes them very reactive. For example, ethyne (acetylene) is an unstable, explosive gas.

1. COMBUSTION

As with the other hydrocarbons, alkynes release large amounts of energy on burning. Thus in the oxyacetylene burner (for cutting and welding metals) the flow of oxygen and acetylene gases can be regulated to give flame temperatures (up to 3000°C) above the melting points of most metals.

2. SATURATION REACTIONS

The most characteristic reactions of alkynes, as with alkenes, are addition reactions, resulting in the removal of triple bonds, e.g.

$$\text{Hydrogenation } CH \equiv CH \xrightarrow{H_2} CH_2 = CH_2 \xrightarrow{H_2} CH_3 - CH_3$$
(normally requires a finely divided metal catalyst)

$$\text{Halogenation } CH \equiv CH \xrightarrow{Br_2} CHBr = CHBr \xrightarrow{Br_2} CHBr_2 - CHBr_2.$$

Under acidic conditions and in the presence of a catalyst ($HgSO_4$), a water molecule can also be added across the triple bond:

$$CH_3 - C \equiv CH + H_2O \rightarrow CH_3 - \overset{\overset{\displaystyle OH}{|}}{C} = \overset{\overset{\displaystyle H}{|}}{CH}.$$

The product contains the *enol group* (a hydroxyl functional group attached to a double-bonded carbon atom). This enol product is unstable and immediately undergoes a rearrangement to give a keto group:

$$CH_3 - \overset{\overset{\displaystyle OH}{|}}{C} = \overset{\overset{\displaystyle H}{|}}{CH} \rightarrow CH_3 - \overset{\overset{\displaystyle O}{\|}}{C} - CH_3 \qquad \text{(where } -\overset{\overset{\displaystyle O}{\|}}{C} - \text{ is}$$
$$\text{propanone} \qquad \text{a keto group)}$$
$$\text{(acetone)}$$

by the migration of a hydrogen atom from the hydroxyl group to the next carbon atom.

In general, the *reversible* reaction:

$$\text{Enol form} \rightleftharpoons \text{Keto form}$$

is called Tautomerism and the two forms are different *Tautomers* of the same compound. In the present example, almost all of the substance exists as the keto tautomer (the equilibrium position is far to the right). However, the enol tautomer may be the more stable form in other compounds, e.g. acetylacetone

$$
\underset{\text{Enol}}{CH_3-\overset{\overset{\displaystyle OH}{|}}{C}=CH-\overset{\overset{\displaystyle O}{\|}}{C}-CH_3} \rightleftharpoons \underset{\text{Keto}}{CH_3-\overset{\overset{\displaystyle O}{\|}}{C}-CH_2-\overset{\overset{\displaystyle O}{\|}}{C}-CH_3.}
$$

If acetylacetone existed as a simple chain molecule, it would take the keto form as in acetone. However, *in the enol form only*, a more stable ring structure can be formed by intramolecular hydrogen bonding and, therefore, 80% of the molecules of acetylacetone exist in the *enol* form—

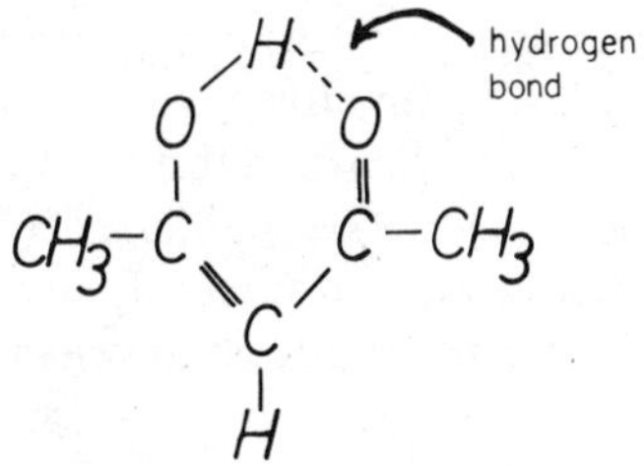

Enol $\rightleftharpoons$ Keto tautomerism is very important in biochemistry. For example, in Chapter 14 we shall see that it is a critical factor in binding together the two strands of the DNA double helix (Figure 14.3).

Important Alkynes

Acetylene is by far the most common and most important alkyne, but there are a few naturally occurring, complex examples, e.g.

$$CH_3-C\equiv C-C\equiv C-C\equiv C-C\equiv C-C\equiv C-CH=CH_2$$
from flowers of certain species of the Compositae (Daisy family).

12.4 THE ARENES (AROMATIC HYDROCARBONS)

The simplest and most important aromatic hydrocarbon is benzene, C_6H_6. In benzene, the carbon atoms are arranged in a ring, and since carbon has a valency of 4, we might expect the carbon atoms to be linked by alternate single and double bonds (I). For simplicity, we normally draw this structure as shown (II).

This suggests that benzene is like an alkene, and, therefore, we would expect it to behave as if it contained three normal double bonds. However, benzene behaves unusually in several ways, for example:

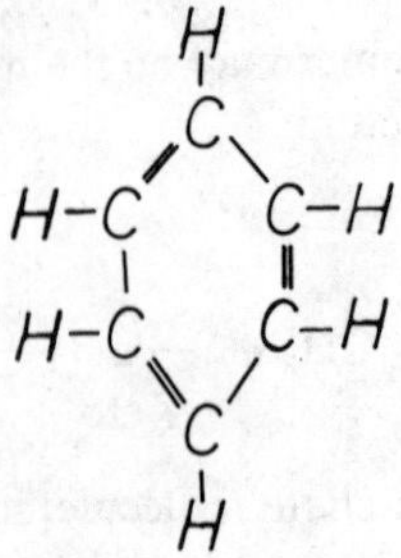

(1) Alkenes readily undergo *addition reactions* resulting in a saturated molecule. Benzene does not; instead, it tends to undergo reactions in which hydrogen atoms are *replaced* by other atoms or groups (*substitution reactions*). Substitution reactions normally proceed only under vigorous conditions (high temperature, high concentration of reagents, presence of a catalyst), e.g. reaction with bromine (test for unsaturation in organic molecules)—

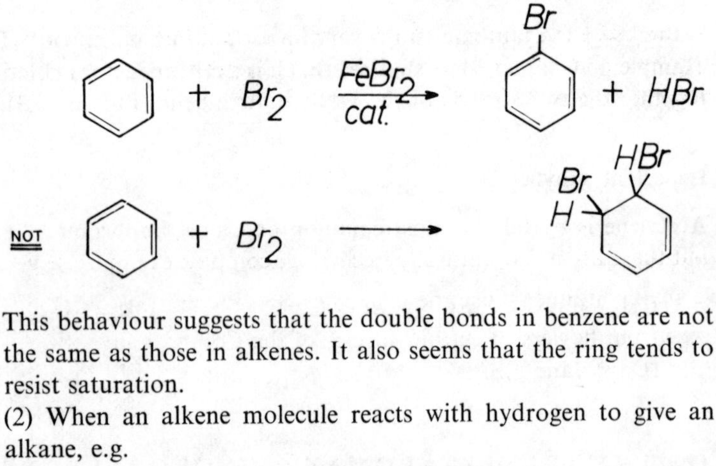

This behaviour suggests that the double bonds in benzene are not the same as those in alkenes. It also seems that the ring tends to resist saturation.

(2) When an alkene molecule reacts with hydrogen to give an alkane, e.g.

$$CH_2{=}CH_2 + H_2 \rightarrow CH_3{-}CH_3,$$

the reaction is exergonic and results in a *more stable* (lower energy) product. Under certain conditions, benzene can also react with hydrogen to give cyclohexane:

$$C_6H_6 + 3H_2 \rightarrow C_6H_{12}.$$

However, the heat of reaction is much less than would be predicted for the saturation of three double bonds, again stressing the chemical stability (i.e. lower energy) of the benzene ring bonds as compared with those in alkenes.

(3) There should be two isomers of 1,2-dimethyl benzene—

i.e. one with a single bond between the carbon atoms carrying methyl groups and one with a double bond. In fact, these two isomers turn out to be the same compound.

The explanation of these apparent anomalies is that the bonds between carbon atoms in the benzene ring are *all the same*, each being intermediate between a single and a double bond so that the valency of carbon is fulfilled. This is shown by the fact that a typical single bond between C atoms is 1·54 Å long, a typical double bond between C atoms is 1·34 Å whereas the C—C bond length in benzene is intermediate, 1·39 Å (1 Å = 10^{-10} m). We can think of the true structure of benzene as being intermediate between the two structures—

Sometimes the intermediate structure is drawn as shown (III), but we shall continue to use the conventional structure (IV)—

III   or   IV

The presence of six intermediate bonds in its ring makes benzene more like an alkane than an alkene. As long as it is intact, the benzene ring is stable and unreactive.

The Naming of Substituted Benzene Compounds

Substituted benzene molecules are named by numbering the carbon atoms clockwise round the ring and using these numbers to indicate the position of attached functional groups —

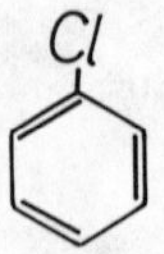

chlorobenzene

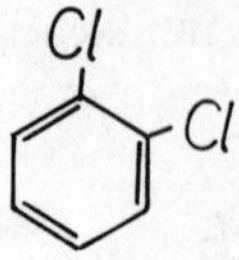

1,2-dichlorobenzene (or ortho- dichlorobenzene)

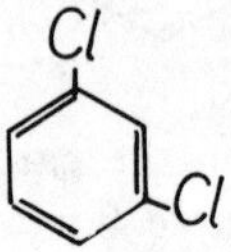

1,3-dichlorobenzene
(or meta-dichlorobenzene)

1,4-dichlorobenzene
(or para-dichlorobenzene)

Chemical Properties and Reactions of Arenes

Aromatic hydrocarbons derive their name (aromatic) from their strong, usually pleasant smells. Most are toxic and several, including benzene, are carcinogenic (cause cancer). Arenes are highly hydrophobic, and benzene is a very effective (but dangerous) solvent for grease. In general, aromatic hydrocarbons are chemically inert, although, as we shall see later, the presence of attached atoms or groups can make them more reactive.

The Importance and Occurrence of Arenes

Unsubstituted arene molecules do not normally occur in living organisms; in fact, we have noted above that these compounds are toxic to humans. On the other hand, there are countless substituted aromatic compounds involved in biochemical pathways; for example, three amino acids necessary for the synthesis of proteins contain benzene rings, although the mammalian body cannot synthesize the benzene ring. Consequently, these three amino acids, Phenylalanine, Tryptophan and Tyrosine are amongst the *essential amino acids*, which must be derived (directly or indirectly) from plants (section 14.2). The unreactivity of aromatic rings also has important consequences in several branches of environmental science; in particular it contributes to the persistence in the soil (resistance to chemical attack) of some aromatic pesticides and to the stability of soil humus.

Pure aromatic hydrocarbons can be synthesized industrially or obtained from petroleum or coal. Various substituted arenes are important in the industrial production of dyes, explosives, plastics, drugs, disinfectants, etc.

Arenes other than Benzene

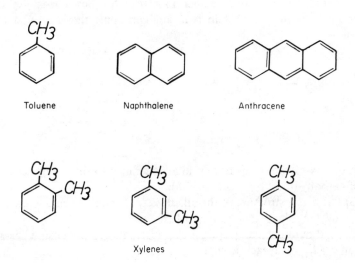

Toluene   Naphthalene   Anthracene

Xylenes

Note that there are aromatic compounds which include atoms other than carbon in the ring. These compounds are called *Heterocyclic e.g.—*

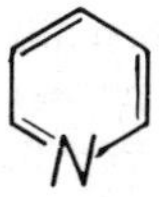

Pyridine 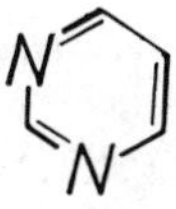   pyrimidine

EXERCISES

(1) We have seen that butane has two structural isomers. Pentane has 3, hexane 5, heptane 9, octane 18, nonane 35 and so on up to $C_{40}H_{82}$

which has an estimated 62, 491, 178, 805, 831 possible structural isomers. Draw all the isomers of pentane, hexane and heptane, and give their systematic names.

(2) Using a 'ball and stick' molecule construction kit, build models of long chain and branched alkane molecules. Notice that five carbon atoms can be linked together to give a flat ring (cyclopentane, whose structure is similar to the furanose rings in carbohydrates) whereas six carbon atoms link together to give a 'puckered ring' (cyclohexane) which can exist in two configurations, the *chair* and the *boat*—

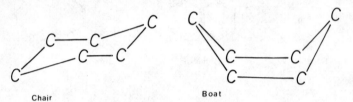

Chair   Boat

These six-carbon rings have similar structures to the pyranose rings of carbohydrates (see section 13.5).

(3) Draw the structure of the alkane 2,2,4-trimethyl pentane used in establishing the octane rating of fuels.

(4) Find out how many straight-chain isomers of pentene and hexene can exist. Name each isomer.

FURTHER READING

Details of the occurrence, composition, refining and useful fractions of petroleum are given in:
1. *Modern Petroleum Technology*, 3rd Edition. Institute of Petroleum (1962).
For a discussion of the influence of oil-spills on marine life:
2. VERNBERG F.J. & VERNBERG W.B. (eds) (1974) *Pollution and Physiology of Marine Organisms*. Academic Press, London.
For a review of the chemistry of air pollutants, including hydrocarbons and airborne lead from petrol:
3. SPEDDING D.J. (1974) *Air Pollution*. Oxford University Press, Oxford.
The many roles that ethene can play in the control of plant growth and development are reviewed in:
4. ABELES F.B. (1973) *Ethylene in Plant Biology*. Academic Press, London.

CHAPTER 13
ORGANIC COMPOUNDS
CONTAINING OXYGEN

In Chapter 12, we considered the four main groups of hydrocarbons, which contain carbon and hydrogen atoms only, but most bio-chemical substances (carbohydrates, lipids etc.) also have oxygen-containing functional groups attached to the hydrocarbon frame-work. In this chapter, we shall look at the properties and reactions associated with these functional groups.

13.1 THE ALCOHOLS

The alcohols are aliphatic (*not* aromatic) compounds containing one or more hydroxyl groups attached to a hydrocarbon chain, which may be straight or branched, saturated or unsaturated. Some simple examples have already been illustrated in section 11.2.

The Naming of Alcohols

Using the systematic method of naming organic molecules, we delete the final -e in the name of the corresponding alkane and substitute -ol. Thus,

$$CH_4 \quad \text{gives} \quad CH_3OH$$
Methane Methanol

and

$$CH_3 \quad \text{gives} \quad CH_3$$

$$CH_3—CH_2CH—CH_2—CH_3 \qquad CH_3—CH_2—CH—CH_2—CH_2—OH$$
3-methyl-pentane   3-methyl-l-pentanol

where the number preceding the basic name (pentanol) indicates the position of the hydroxyl group. We shall not normally require to name more complicated alcohols (highly branched and unsaturated).

The carbon atom to which the hydroxyl group is attached may be called the α-carbon atom and it is possible to classify alcohols according to the number of alkyl groups (written as R groups) attached to the α-carbon, i.e.

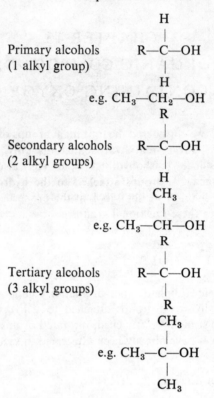

Primary alcohols
(1 alkyl group)

$$\begin{array}{c} H \\ | \\ R-C-OH \\ | \\ H \end{array}$$

e.g. CH_3-CH_2-OH

Secondary alcohols
(2 alkyl groups)

$$\begin{array}{c} R \\ | \\ R-C-OH \\ | \\ H \end{array}$$

e.g. $CH_3-CH-OH$ with CH_3

Tertiary alcohols
(3 alkyl groups)

$$\begin{array}{c} R \\ | \\ R-C-OH \\ | \\ R \end{array}$$

e.g. CH_3-C-OH with CH_3 above and CH_3 below

TABLE 13.1. Properties of some straight-chain alcohols.

Name	Formula	Molecular weight	m.p. °C	b.p. °C	Water solubility (g 100 ml^{-1})
Methanol	CH_3OH	32	-94	65	*
Ethanol	C_2H_5OH	46	-117	79	*
1-propanol	C_3H_7OH	60	-127	97	*
1-butanol	C_4H_9OH	74	-90	117	8·0
1-pentanol	$C_5H_{11}OH$	88	-79	137	2·2
1-hexanol	$C_6H_{13}OH$	102	-47	158	0·6
1-heptanol	$C_7H_{15}OH$	116	-34	176	0·1
1-octanol	$C_8H_{17}OH$	130	-17	195	0·04
1-decanol	$C_{10}H_{21}OH$	158	7	229	0·004

* Soluble in all proportions of water to alcohol.

Properties and Reactions of Alcohols

1. PHYSICAL PROPERTIES

If we compare the melting and boiling points of alcohols (Table 13.1) with those of the corresponding alkanes (Table 12.1), we find that the alcohols have much higher values, although the differences diminish as the molecules become longer. One result of these higher values is that there are no gaseous alcohols.

The relatively high temperatures for changes of state and the very high water solubilities of the lower alcohols (C1 to C5) are due to the formation of strong hydrogen bonds between the hydroxyl groups of alcohol molecules and also between hydroxyl groups and water molecules, e.g.

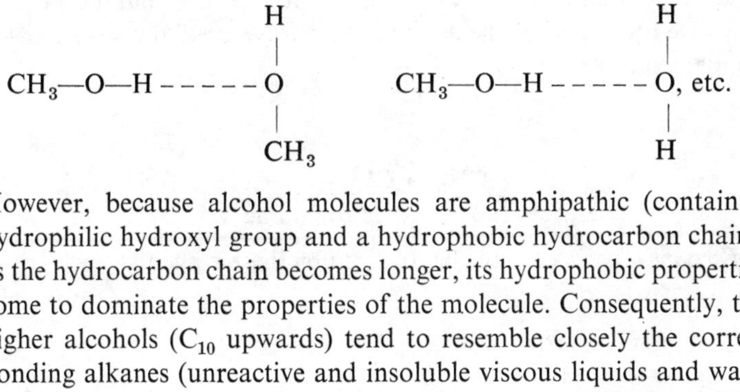

$$
\begin{array}{cc}
\text{H} & \text{H} \\
| & | \\
\text{CH}_3\text{—O—H} - - - - - \text{O} \qquad \text{CH}_3\text{—O—H} - - - - - \text{O, etc.} \\
| & | \\
\text{CH}_3 & \text{H}
\end{array}
$$

However, because alcohol molecules are amphipathic (contain a hydrophilic hydroxyl group and a hydrophobic hydrocarbon chain), as the hydrocarbon chain becomes longer, its hydrophobic properties come to dominate the properties of the molecule. Consequently, the higher alcohols (C_{10} upwards) tend to resemble closely the corresponding alkanes (unreactive and insoluble viscous liquids and waxy solids).

2. ACID-BASE PROPERTIES

Alcohols are *very* weak acids. For the ionization

$$ROH + H_2O \rightleftharpoons RO^- + H_3O^+$$

K_a values vary from 10^{-13} to 10^{-18} depending upon the length and nature of R. Under certain vigorous conditions, salts of alcohols can be formed, e.g. the reaction of pure alcohols with alkali metals:

$$2ROH + 2Na \rightarrow 2Na^+RO^- + H_2 \uparrow$$

(compare with the reaction between water and sodium in section 7.1).

3. Redox Reactions

These reactions are used in the laboratory to distinguish between primary, secondary and tertiary alcohols.

Primary alcohols are oxidized by strong oxidizing agents (e.g. potassium dichromate or potassium permanganate) to give, first, aldehydes and, subsequently, carboxylic acids, e.g.

$$\underset{\text{Ethanol}}{CH_3CH_2OH} \rightarrow \underset{\substack{\text{Ethanal} \\ \text{(Acetaldehyde)}}}{CH_3\overset{\overset{\displaystyle O}{\|}}{C}-H} \rightarrow \underset{\substack{\text{Ethanoic acid} \\ \text{(Acetic Acid).}}}{CH_3\overset{\overset{\displaystyle O}{\|}}{C}-OH}$$

A practical example of this reaction is the production of sour wine or beer (containing acetic acid) when a fermentation mixture becomes contaminated with ethanol-oxidizing micro-organisms.

Secondary alcohols can be oxidized to ketones, but no further (because formation of the acid would involve breaking strong C—C bonds), e.g.

$$\underset{\text{2-propanol}}{CH_3-\overset{\overset{\displaystyle OH}{|}}{CH}-CH_3} \rightarrow \underset{\text{Propanone}}{CH_3-\overset{\overset{\displaystyle O}{\|}}{C}-CH_3} \nrightarrow \text{Acid}.$$

Tertiary alcohols cannot be oxidized by these reagents.

4. Esterification

Alcohols react with carboxylic acids to give *esters*. The general reaction may be written:

$$R-OH + R^1-\overset{\overset{\displaystyle O}{\|}}{C}-OH \rightarrow R^1-\overset{\overset{\displaystyle O}{\|}}{C}-OR + H_2O.$$

For example, the reaction of ethanol with ethanoic (acetic) acid

$$C_2H_5-OH + CH_3-\overset{\overset{\displaystyle O}{\|}}{C}-OH \rightarrow CH_3-\overset{\overset{\displaystyle O}{\|}}{C}-OC_2H_5 + H_2O$$

yields the fruity-smelling ester ethyl acetate which is a versatile solvent for organic substances (e.g. in glues and varnishes, and in dry cleaning textiles). Esters play a part in many biochemical pathways, notably in energy and lipid metabolism, as well as acting, due to their strong pleasant smells, as chemical attractants in nature (i.e. to ripe fruit, and between insects). Their chemistry is treated in more detail in section 13.6.

5. FORMATION OF HEMIACETALS AND CONDENSATION OF ALCOHOLS

Two further types of reaction are important in the structure and chemistry of carbohydrates (sugars, starch, cellulose, etc.) which are polyhydroxy alcohols containing aldehyde and ketone groups (see sections 13.4 and 13.5). Firstly, an alcohol can react with an aldehyde to give a hemiacetal:

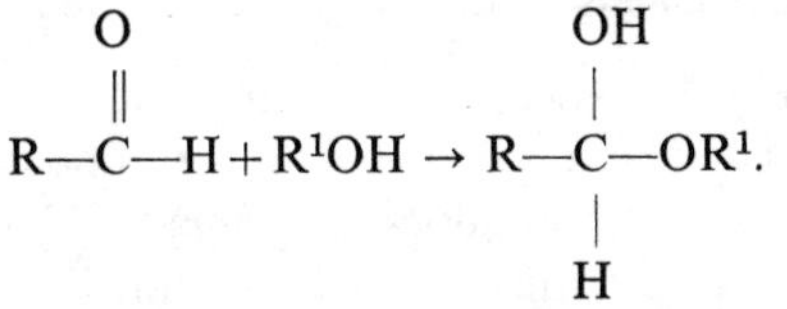

$$\underset{\text{Aldehyde}}{\overset{\overset{\textstyle O}{\|}}{R-C-H}} + \underset{\text{Alcohol}}{R^1 OH} \rightarrow \underset{\text{Hemiacetal}}{\overset{\overset{\textstyle OH}{|}}{\underset{\overset{|}{H}}{R-C-OR^1.}}}$$

A similar reaction occurs with ketones. Where both aldehyde and alcohol functional groups are on the *same* molecule, this (intra-molecular) reaction gives a ring structure (e.g. the pyranose rings of glucose, section 13.5).

As a result of the second reaction type, the condensation, or polymerization, of two (or more) alcohol molecules,

$$R-OH + HO-R^1 \rightarrow R-O-R^1 + H_2O,$$

many single monosaccharide units can be joined together to give long polysaccharide chains (section 13.5).

Ethanol (Ethyl Alcohol)

Ethanol is very useful in the laboratory as a hydrophilic solvent, as a dehydrating agent in preparing material for microscopy and as a preservative for plant and animal specimens. It is also the important ingredient in alcoholic drinks—beer (about 4%), wine (10–20%) and spirits (about 40%).

Three main types of ethanol are used in the laboratory:

Absolute   —100% ethanol;

Rectified   —95% ethanol (the remaining 5% including water and benzene);

Denatured—as for rectified but containing methanol, 2-propanol
or benzene to make it unfit for human consumption
e.g. methylated spirits contains methanol.

The concentration of ethanol in alcoholic drinks is normally expressed as *degrees proof* (approximately twice the percentage composition).

Ethanol is a common end-product of anaerobic respiration (fermentation) in higher plants (as in other organisms); its effects on cell membranes are thought to be at least partly responsible for the damaging effects of flooding on sensitive plant species.

Other Important Alcohols

Methanol, CH_3OH (wood alcohol), may be obtained by heating certain hardwoods and condensing the vapour produced. It is highly toxic, causing blindness in humans. Methanol is used as a solvent, for denaturing ethanol and in the synthesis of formaldehyde.

Ethylene glycol, $CH_2—CH_2$, is a dihydroxy alcohol used as a sol-
               | |
              OH OH
vent, as anti-freeze in vehicle radiators and as a coolant liquid for refrigerators.

Glycerol, $CH_2—CH—CH_2$ (glycerine), is a sweet-tasting, viscous
          | | |
        OH OH OH
trihydroxy alcohol. It is an important constituent of lipids (section 13.6) and is a valuable by-product of soap manufacture. Due to its hygroscopic nature, glycerol is mixed with tobacco to keep it moist.

Monosaccharides are polyhydroxy alcohols, e.g. glucose

$$\underset{\displaystyle H\quad\ OH\ \ H\ \ \ H\ \ \ \ H}{\overset{\displaystyle O\quad OH\ \ H\ \ OH\ OH\ OH}{H—C—C—C—C—C—C—H}}$$

Long-chain alcohols are used as detergents.

Many other biochemical molecules contain the hydroxyl group, e.g. Vitamin A1 (section 12.2).

13.2 THE PHENOLS

The phenols are similar to alcohols except that the hydroxyl functional group is attached directly to an aromatic ring. The simplest member of the series is phenol (carbolic acid) (I) —

I

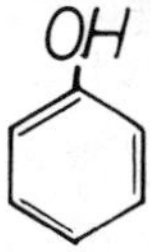

II

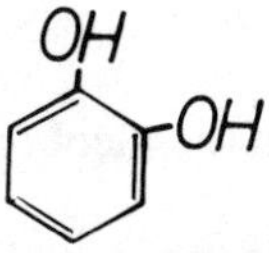

More complicated members may be named using the procedure outlined in section 12.4. However, aromatic compounds are commonly given special, unsystematic names. One example (II), is normally called Catechol, although correctly it is 1,2-dihydroxy benzene or orthodihydroxy benzene. Other examples are given at the end of this section.

Properties and Reactions of Phenols

Although phenols behave like aliphatic alcohols in a few reactions (e.g. esterification with carboxylic acids), the influence of the aromatic ring makes the chemistry of the hydroxyl group generally very different in phenols.

1. ACID/BASE PROPERTIES

Because the benzene ring tends to *attract electrons*, the hydrogen atom in a phenol hydroxyl group is not held as tightly as the hydrogen atom in an aliphatic alcohol hydroxyl group —

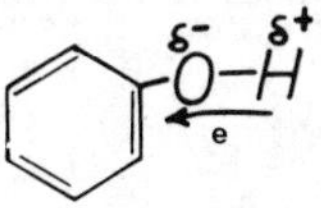

Consequently the equilibrium position in the reaction —

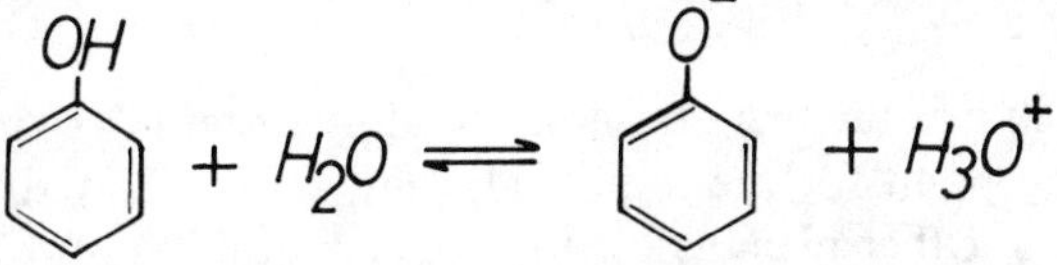

is further to the *right* than in a reaction:

$$R—OH + H_2O \rightarrow RO^- + H_3O^+$$

with the result that phenols are more acidic than alcohols, as shown by the following K_a values:

Phenol	10^{-10}	Compare with the values for acids in
Aliphatic alcohols	10^{-13} to 10^{-18}	Table 8.2.

Phenol is normally considered to be an acid and readily forms salts with alkalis and alkali metals. Alcohols are not acids.

2. REDOX REACTIONS

Phenols are more readily oxidized than alcohols e.g.—

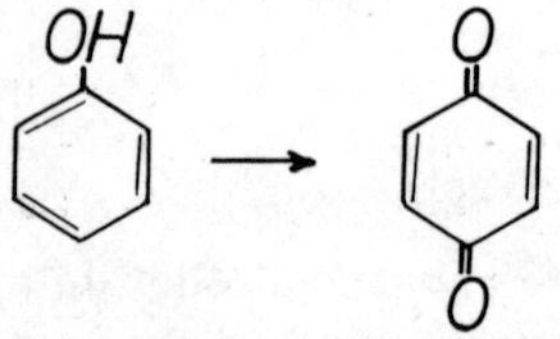

p – benzoquinone

The products of oxidation are normally highly coloured *quinones* (cyclic diketones). This type of redox reaction, which does not require strong oxidizing agents, is employed in photographic developing, where the silver bromide on an exposed film is precipitated as silver by the redox reaction—

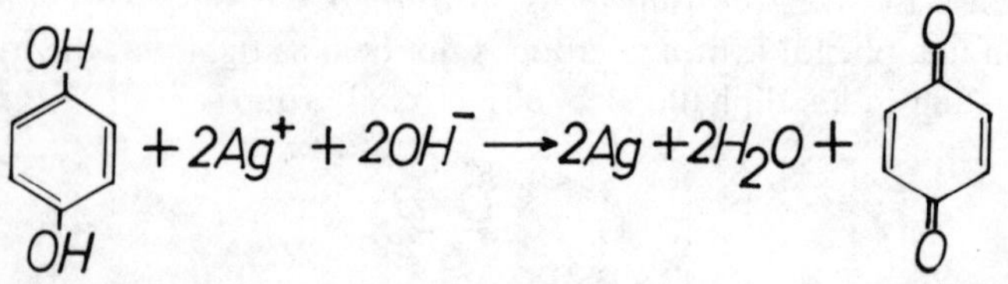

Hydroquinone

Uses of Phenols

Phenol itself was the first antiseptic, used by Lister in 1865. It is also the starting point for the industrial synthesis of many drugs, dyes and plastics. Other phenols used in the chemical industry include—

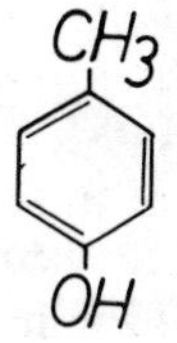

p-cresol 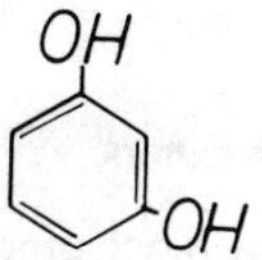   Resorcinol 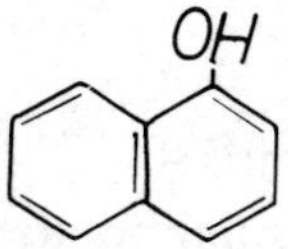   1-naphthol (α-naphthol)

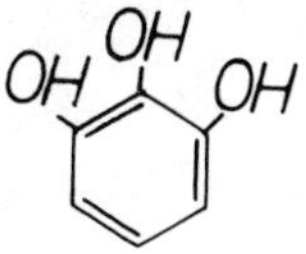

Pyrogallol 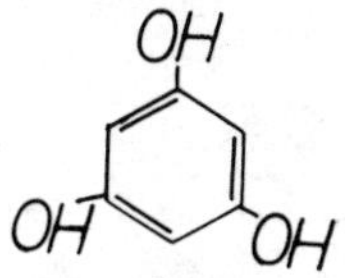   Phloroglucinol

13.3 THE ETHERS

In an ether molecule, two aliphatic groups, two aromatic groups (Ar), or one of each type are linked together through a single oxygen atom

i.e.   R—O—R,   Ar—O—Ar   R—O—Ar.

e.g. CH_3—O—C_2H_5,
          methyl ethyl ether

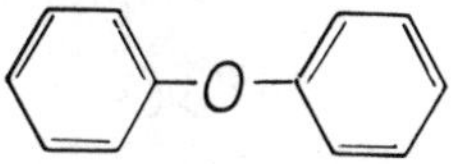

Diphenyl ether

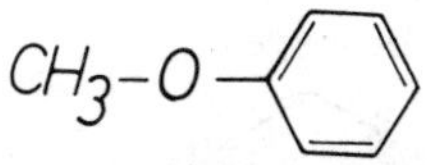

Methyl phenyl ether
(Anisole)

As indicated, ethers are named simply by listing the two groups attached to the oxygen atom followed by the word ether. (Note the name *phenyl* for an attached benzene ring.)

Properties and Reactions of Ethers

The outstanding property of ethers is their unreactivity and, because of this, they are commonly used as inert solvents for organic reactions. However, ethers can react explosively in the presence of light and oxygen and for this reason they are normally kept in brown bottles to exclude light.

The Occurrence and Uses of Ethers

The ether link occurs in many biochemical molecules, for example in, and between, the rings of carbohydrates (section 13.5) and in natural products such as —

Vanillin
(vanilla)

Diethyl ether, $C_2H_5O\ C_2H_5$, sometimes simply called ether, is probably the most important of the series because of its use as a solvent and anaesthetic. However, diethyl ether must be handled with the greatest care; the high density of its vapour (greater than air) can result in the accumulation of pockets of inflammable gas, for example, in laboratory drains.

Other ethers, important as solvents and in the synthesis of organic compounds, include—

α‑pyran   Furan   Dioxane

Note that the names *pyranose* and *furanose* for 6- and 5-membered monosaccharide rings originate from pyran and furan.

13.4 ALDEHYDES AND KETONES

Aldehyde and ketone molecules contain the carbonyl functional group,

In aldehydes, the group is at the end of a chain, e.g. Ethanal (acetaldehyde)

$$CH_3-\overset{\overset{\displaystyle O}{\|}}{C}-H$$

whereas in ketones, it occurs within the chain, e.g. Propanone (acetone, dimethyl ketone)

$$CH_3-\overset{\overset{\displaystyle O}{\|}}{C}-CH_3.$$

We shall see (section 13.6) that carboxylic acids and acid derivatives also carry the carbonyl group, but in close association with another functional group.

Naming Aldehydes and Ketones

Aldehydes—using the systematic method, we delete the final -e in the name of the corresponding alkane and substitute -al.

Thus C_2H_6 gives CH_3CHO
              Ethane   Ethanal

and
$$CH_3CH_2-\overset{\overset{\displaystyle CH_3}{|}}{CH}-CH_2-CH_3$$
3-methyl-pentane

gives
$$CH_3CH_2-\overset{\overset{\displaystyle CH_3}{|}}{CH}-CH_2-CHO$$
3-methyl-pentanal

where the aldehyde-group-carbon atom is always numbered C1.

Ketones—using the systematic method we delete the final -e of the name of the corresponding alkane and substitute -one, e.g.

$CH_3CH_2CH_2CH_3$ gives $CH_3\overset{\overset{\displaystyle O}{\|}}{C}CH_2CH_3$
Butane   2-butanone

where the position of the carbonyl group is indicated by the number preceding the main name, e.g.

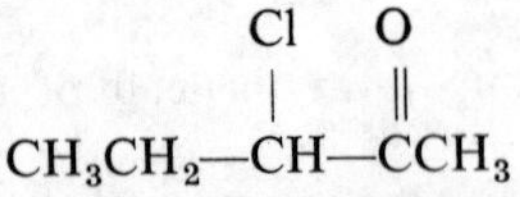

$$CH_3CH_2—\overset{\overset{\displaystyle Cl}{\displaystyle |}}{CH}—\overset{\overset{\displaystyle O}{\displaystyle \|}}{C}CH_3$$
3-chloro-2-pentanone.

Ketones may also be named like ethers, although this is a more clumsy method, e.g.

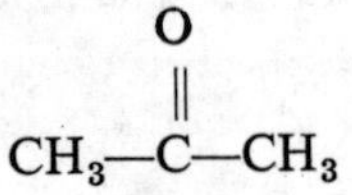

$$CH_3—\overset{\overset{\displaystyle O}{\displaystyle \|}}{C}—CH_3 \qquad \text{Dimethyl ketone}$$

TABLE 13.2. Properties of some aldehydes and ketones.

Name	Formula	Molecular weight	m.p. °C	b.p. °C	Water solubility (g 100 ml^{-1})
Aldehydes					
Methanal (Formaldehyde)	HCHO	30	−92	−21	Very soluble
Ethanal (Acetaldehyde)	CH_3CHO	44	−121	21	*
Propanal	C_2H_5CHO	58	−81	49	16
Butanal	C_3H_7CHO	72	−99	76	4
Pentanal	C_4H_9CHO	86	−92	103	Slightly soluble
Ketones					
Propanone (Acetone)	CH_3COCH_3	58	−95	56	*
2-butanone	$CH_3CO\,C_2H_5$	72	−86	80	33
2-pentanone	$CH_3CO\,C_3H_7$	86	−78	102	6
3-pentanone	$C_2H_5CO\,C_2H_5$	86	−40	102	5
2-hexanone	$CH_3CO\,C_4H_9$	100	−57	128	1·6

* Soluble in all proportions.

Properties and Reactions of Aldehydes and Ketones

As with most organic compounds, the physical properties of aldehydes and ketones are a compromise between the hydrophobic and hydrophilic parts of the molecule (see Table 13.2). Aldehydes and ketones (like esters) tend to have pleasant smells and flavours (e.g. Vanillin—section 13.3—which is an aldehyde as well as an ether).

Some of the chemical properties of aldehydes and ketones have been described in previous sections, i.e. Keto/Enol tautomerism (12.3) and the formation of Hemiacetals (13.1 and 13.5).

1. REDOX REACTIONS

Alipathic aldehydes can be oxidized to the corresponding carboxylic acid, using either Fehling's or Tollen's reagent (see the exercise section at the end of the chapter for details), whereas aliphatic ketones can not. This is an important method for discriminating between the two types of compound and is the basis for the classification of monosaccharides into reducing (aldehyde) and non-reducing (ketone) sugars, e.g.

$$
\begin{array}{ccc}
O & & O \\
\| & & \| \\
CH_3\text{—}C\text{—}H & \rightarrow & CH_3\text{—}C\text{—}OH \\
\text{Ethanal} & & \text{Ethanoic acid}
\end{array}
$$

$$
\begin{array}{c}
O \\
\| \\
CH_3\text{—}C\text{—}CH_3 \;\;\cancel{\rightarrow}\;\; \text{Acid} \\
\text{Propanone}
\end{array}
$$

2. CONDENSATION BY THE ALDOL REACTION

Under alkaline conditions, two molecules of ethanal can *condense* together:

$$
\begin{array}{ccc}
O & & OH \quad O \\
\| & & | \quad\;\; \| \\
2CH_3C\text{—}H & \rightarrow & CH_3CHCH_2C\text{—}H
\end{array}
$$

to give 3-hydroxy-butanal. This reaction is a simple example of an *Aldol Condensation*, and it can proceed further to give a long chain of carbon atoms. In a slightly different form, this reaction is involved in the biosynthesis of long-chain fatty (carboxylic) acids from 2-carbon molecules.

3. POLYMERIZATION REACTIONS

In aqueous solution, methanal (formaldehyde) tends to polymerize to give a long-chain compound (paraformaldehyde) which is the active ingredient of the biological preservative, formalin, i.e.

$$\begin{array}{c} O \\ \parallel \\ n\text{H—C—H} + H_2O \end{array} \rightarrow HO\text{—CH}_2\text{—(OCH}_2)_{\overline{n-2}}O\text{—CH}_2OH.$$

On heating, the gaseous monomer methanal is again released. Under other conditions, a trimer can be formed—

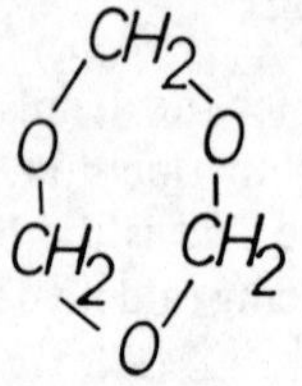

Trioxymethylene

Similar polymerization reactions occur with ethanal giving a trimer or a tetramer—

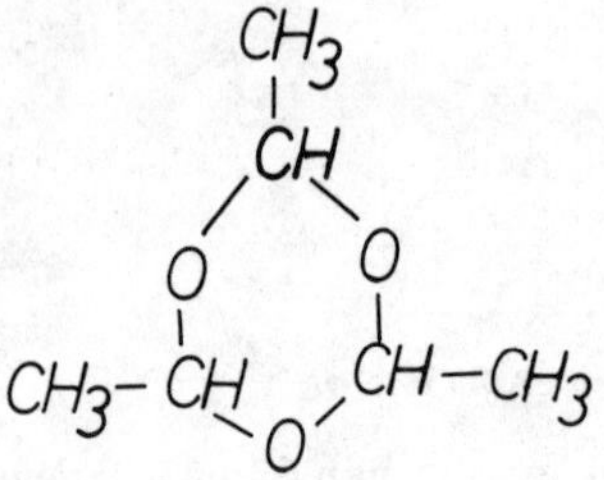

Paraldehyde

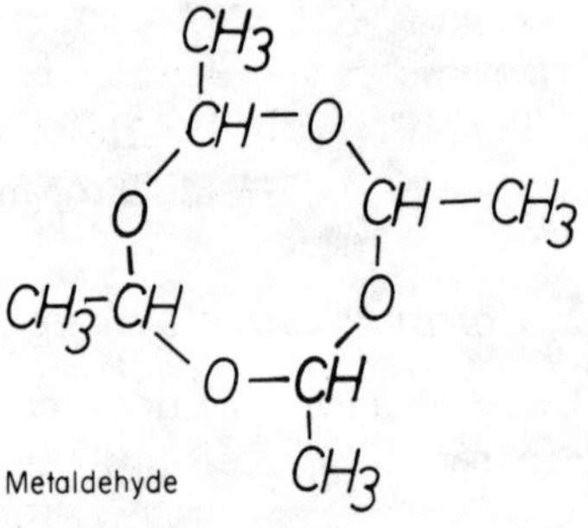

Metaldehyde

Paraldehyde is a sleep-inducing drug whereas metaldehyde is a slug and snail poison (molluscicide).

Important Aldehydes and Ketones

Methanal (formaldehyde), in the form of formalin (37% methanal in water) is used as a disinfectant and preservative. It is also used in the tanning of hides, the synthesis of plastics and drugs, as a fumigant and in the manufacture of mirrors (see Tollen's test). Methanal is a gas at room temperature.

Propanone (acetone) is a useful solvent. It is a normal intermediate product of the breakdown of lipids in humans and is subsequently oxidized to CO_2 and H_2O. In diabetes, acetone is produced in large quantities in the body and may be smelt on the breath. High ketone concentrations also occur in the blood of cattle as a result of perinatal diseases.

Long-chain aldehydes are used in perfumes.

Monosaccharide molecules carry an aldehyde or ketone group in addition to several hydroxyl groups.

Aromatic Ketones—The Quinones

As we saw in section 13.2, phenols are readily oxidized to quinones, cyclic diketones—

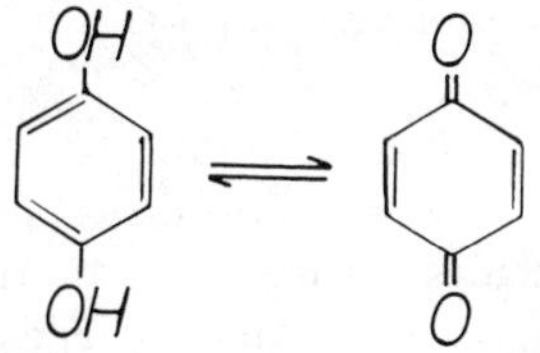

Quinones do not resemble aliphatic ketones in their properties or reactions. Their most characteristic reactions are reversible redox reactions (see above) which result in dramatic colour changes.

The strong colours of quinones are used in dyeing. For example

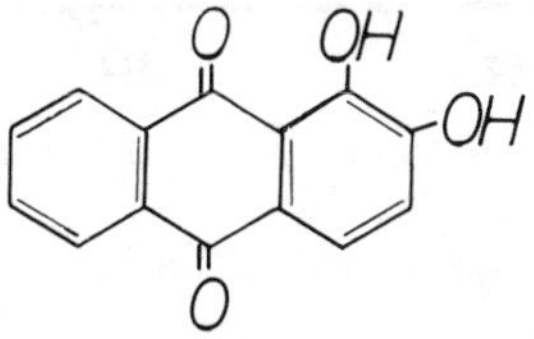

is the structure of Alizarin, the 'Turkey Red' dye extracted from the madder plant. The quinone structure occurs in many other natural molecules; e.g.—

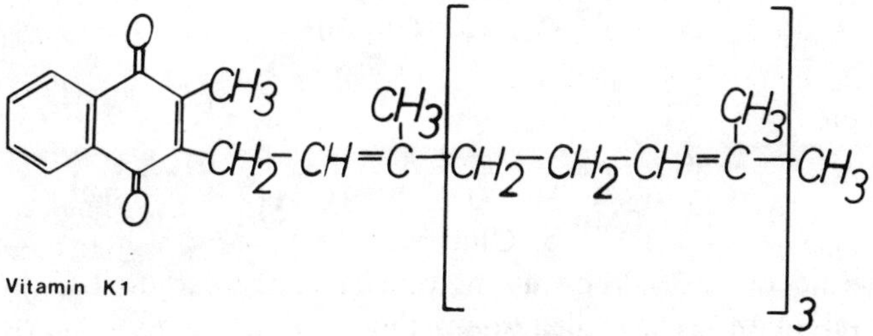

Vitamin K1

13.5 CARBOHYDRATES

Carbohydrates can be defined formally as polyhydroxy aldehydes or ketones (or substances which can be hydrolysed to give polyhydroxy aldehydes or ketones, thus including compounds like phosphorylated sugars and amino sugars). They are particularly important in supplying energy for growth and metabolism, and in providing structural support in plants. The basic formula of all carbohydrates is $(CH_2O)_n$ where n can vary from 3 up to many hundreds or thousands. Whatever the size of a carbohydrate molecule, it is made up of simple units called monosaccharide molecules.

The Monosaccharides

Monosaccharide molecules normally contain from two to six carbon atoms to which are attached one carbonyl and two or more hydroxyl groups.

Examples (Note the system of numbering C atoms which is the same as that for aldehydes and ketones.)

$$
\begin{array}{cc}
\underset{\displaystyle H-C_1-C_2-C_3-H}{\overset{\displaystyle \overset{O}{\|} \quad \overset{OH}{|} \quad \overset{OH}{|}}{}} \\[2mm]
\quad \overset{|}{H} \quad \overset{|}{H} \\
\text{1. Glyceraldehyde (C3)}
\end{array}
\qquad
\begin{array}{cc}
\underset{\displaystyle H-C_1-C_2-C_3-H}{\overset{\displaystyle \overset{OH}{|} \quad \overset{O}{\|} \quad \overset{OH}{|}}{}} \\[2mm]
\quad \overset{|}{H} \qquad \overset{|}{H} \\
\text{2. Dihydroxyacetone (C3)}
\end{array}
$$

$$
\begin{array}{c}
\overset{O}{\|} \quad \overset{OH}{|} \quad \overset{OH}{|} \quad \overset{OH}{|} \quad \overset{OH}{|} \\
H-C_1-C_2-C_3-C_4-C_5-H \\
\quad \overset{|}{H} \quad \overset{|}{H} \quad \overset{|}{H} \quad \overset{|}{H} \\
\text{3. Ribose (C5)}
\end{array}
\qquad
\begin{array}{c}
\overset{OH}{|} \quad \overset{O}{\|} \quad \overset{H}{|} \quad \overset{OH}{|} \quad \overset{OH}{|} \quad \overset{OH}{|} \\
H-C_1-C_2-C_3-C_4-C_5-C_6- \\
\quad \overset{|}{H} \qquad \overset{|}{OH} \quad \overset{|}{H} \quad \overset{|}{H} \quad \overset{|}{H} \\
\text{4. Fructose (6C)}
\end{array}
$$

$$
\begin{array}{c}
\overset{O}{\|} \quad \overset{OH}{|} \quad \overset{H}{|} \quad \overset{OH}{|} \quad \overset{OH}{|} \quad \overset{OH}{|} \\
H-C_1-C_2-C_3-C_4-C_5-C_6-H \\
\quad \overset{|}{H} \quad \overset{|}{OH} \quad \overset{|}{H} \quad \overset{|}{H} \quad \overset{|}{H} \\
\text{5. Glucose (6C).} \qquad\qquad H
\end{array}
$$

Molecules with 3 carbon atoms (including the carbonyl C atom, examples 1 and 2) are called *trioses*, those with 5 (example 3) *pentoses* and those with 6 (examples 4 and 5), *hexoses*. Monosaccharides can be further classified into *reducing sugars*, carrying an aldehyde group (*aldoses*) and *non-reducing sugars*, carrying a ketone group (*ketoses*). Reducing sugars give positive Fehling's and Tollen's tests, whereas non-reducing do not (section 13.4).

Most of the properties of monosaccharides can be predicted from those of alcohols and carbonyl compounds. In particular, they are highly soluble in water due to the presence of several hydrophilic groups. Many monosaccharides and disaccharides have sweet tastes.

Ring Structures of Monosaccharides

We have drawn our examples of pentose and hexose molecules as straight chains. However, these molecules normally exist in ring form due to an intramolecular hemiacetal reaction (section 13.1). For example, in glucose, instead of a reaction between an aldehyde group on one molecule and an alcohol on another:

$$R-OH + R^1-\overset{\overset{O}{\|}}{C}-H \rightarrow R-O-\overset{\overset{OH}{|}}{\underset{\underset{H}{|}}{C}}-R^1$$

the reaction occurs between the two ends (C1 and C5) of the same molecule—

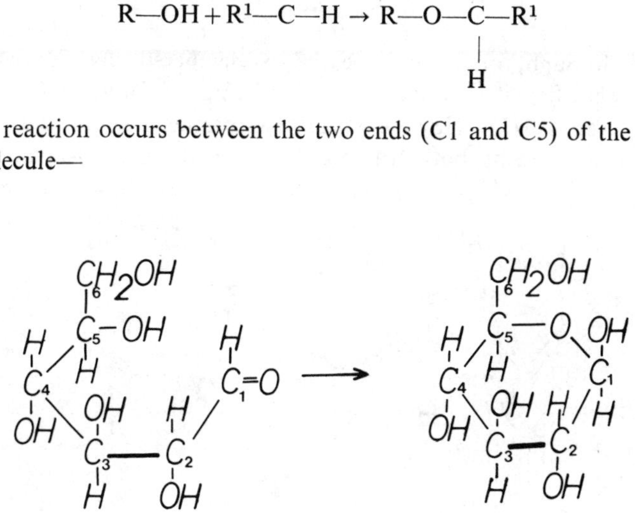

β-form

to give a six-membered *pyranose* ring. Glucose normally exists in the pyranose ring form but the ether linkage in the ring can break readily to give the chain form, when necessary.

Two important aspects of this representation of the pyranose ring should be stressed:

(a) The drawing indicates that the OH group on C1 (as well as H on C2, OH on C3, H on C4 and CH_2OH on C5) is *above* the plane of the ring, whereas the H atom on C1 (as well as the other groups of the molecule) is held below the plane of the ring. This is the β form of the pyranose ring. The hemiacetal reaction can also give the α form in which the hydroxyl group at C_1 is held *below* the ring—

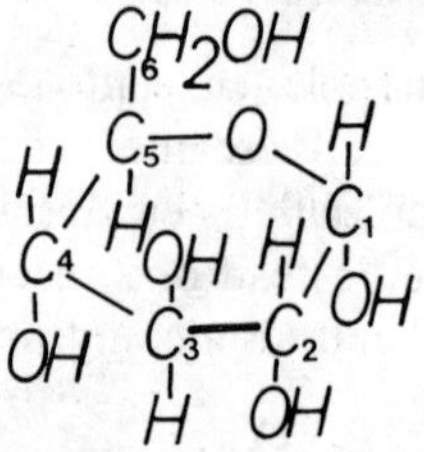

α-form

(b) Although, for convenience, the six-membered ring is drawn as if it were flat, it is in fact a puckered ring, as found in Chapter 12, exercise (2). However, the five-membered furanose ring form adopted by both fructose (6C) and ribose (5C) is planar—

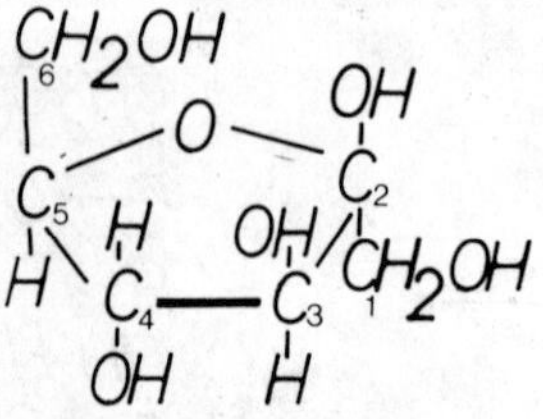

Fructose

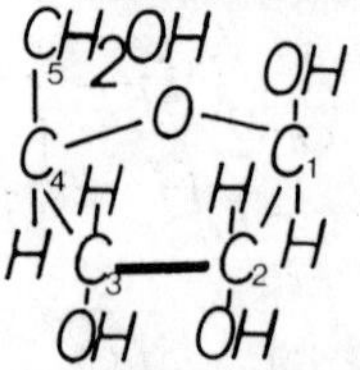

Ribose (β)

(In future diagrams, the C atoms will be omitted, for clarity.)

Although they can form pyranose rings, pentoses like ribose tend to exist in the furanose form. Amongst hexoses, glucose tends to adopt the six-membered, and fructose the five-membered ring.

Importance of Monosaccharides

Glucose is the most commonly occurring monosaccharide (in fruit, starch, cellulose, glycogen, etc.). It is enormously important in biochemistry.

Fructose occurs in fruits and, with glucose, in sucrose.

Ribose molecules are components of RNA and DNA (section 14.4 and Figure 14.3).

The Disaccharides

It was noted in section 13.1, that two alcohol molecules can condense together as follows:

$$R—OH + HO—R^1 \rightarrow R—O—R^1 + H_2O.$$

A similar reaction between hydroxyl groups on two adjacent monosaccharide molecules can lead to the synthesis of a disaccharide molecule. Since monosaccharide molecules are *polyhydroxy* alcohols, there are many possible ways in which the two molecules could link together. However, in nature, the link is normally formed between the C1 of one molecule and the C2, C4 or C6 on the other.

For example, two glucose molecules (α form) can condense together to give the disaccharide, *maltose —*

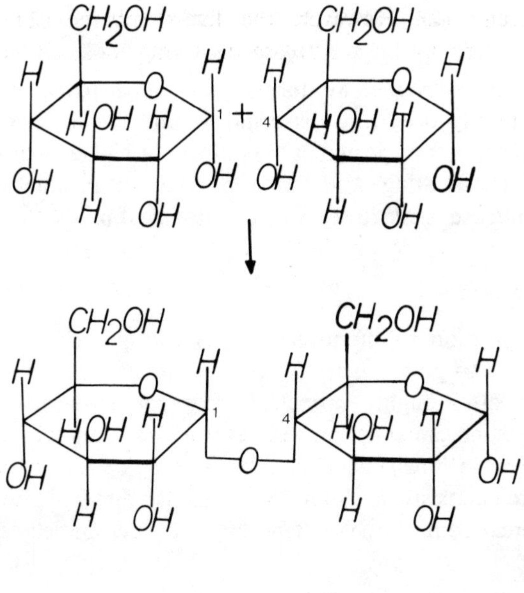

Because the left hand molecule is in the α form and the ether linkage is formed between C1 on the left molecule and C4 on the right, the linkage is called $\alpha(1 \rightarrow 4)$. Similarly an $\alpha(1 \rightarrow 2)$ linkage between molecules of glucose and fructose, gives sucrose —

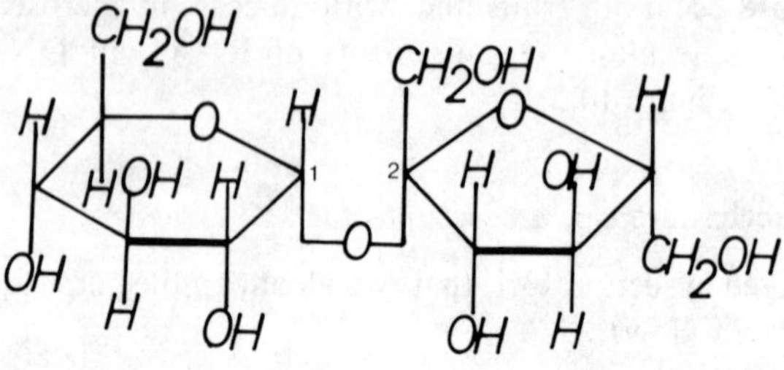

These ether or *glycosidic linkages* between rings can be broken by enzymes or by *acid hydrolysis* to give back the original monosaccharide molecules.

From this brief study of the chemistry of disaccharides, two important conclusions can be drawn:

    (a) in ring closure and in the glycosidic linkage we have important exceptions to the rule that ethers are unreactive, and

    (b) *hydrolysis* reactions are the reverse of *condensation* reactions in carbohydrates (and in lipids and proteins), e.g.

$$\text{glucose} + \text{glucose} \underset{\text{hydrolysis}}{\overset{\text{condensation}}{\rightleftharpoons}} \text{maltose} + H_2O.$$

Sucrose, the substance we normally call sugar (cane sugar), is the most commonly occurring disaccharide and is the form in which photosynthate is transported in the phloem of higher plants. Maltose occurs in germinating seeds as a breakdown product of starch. Lactose (glucose + galactose) is a constituent of milk.

Polysaccharides

The condensation of monosaccharides can proceed further than disaccharides to give *polysaccharides*—complex, long-chain, macromolecules, often highly branched. The condensation of glucose molecules alone can give four different polysaccharides which are of great interest in biochemistry:

    (a) Starch consists of about 25% amylose and 75% amylopectin
      (in cereal grain starch), where *Amylose* macromolecules are long

unbranched chains of glucose units bonded together by $\alpha(1-4)$ linkages, as in maltose, to give molecular weights of 4000 to 50 000; and *Amylopectin* macromolecules can be thought of as branching networks of amylose chains linked together by $\alpha(1-6)$ and $\alpha(1-4)$ glycosidic linkages. They have MW values greater than 500 000.

Starch, which gives a characteristic blue/black colour with iodine, is the form in which carbohydrate is stored in most plant species, particularly in seeds and tubers. Exceptions include certain members of the Compositae and Gramineae which store Inulin (a straight-chain polymer of 30–35 molecules of fructose terminated by a sucrose unit).

(b) Cellulose macromolecules are long chains of glucose units bonded together by $\beta(1-4)$ linkages to give molecular weights of 100 000 to 500 000. Plant cell walls are made of cellulose, impregnated with other materials such as lignin. Important cellulose products include paper and cotton.

Cellulose is much more resistant to hydrolysis than is starch, and obviously much less digestible by humans and livestock; indeed, ruminants rely on bacteria in the rumen to break down cellulose. Since, apart from differences in chain length, cellulose and starch differ principally in the configuration of the glycosidic linkage between glucose units (i.e. $\alpha(1-4)$ compared with $\beta(1-4)$), the higher resistance to hydrolysis in cellulose must be due to the greater chemical stability of the $\beta(1-4)$ linkage.

(c) Dextrins are formed when some of the glycosidic linkages in starch are broken by acid hydrolysis to give macromolecules similar to amylose and amylopectin but of lower molecular weight. Dextrins are used in glues and as food additives ('thickening' agents).

(d) Glycogen is the form in which carbohydrate is stored by animals. Its structure is very similar to amylopectin but with molecular weights of several million.

In nature, there exists a wide range of polysaccharides of glucose and other monosaccharides, commonly in association with substituted molecules, like amino sugars (e.g. gums and pectins in plants; agar from seaweed; chitin in the exoskeleton of insects, and lubricants, such as hyaluronic acid, in vertebrate joints). Due to the presence of many hydroxyl groups, polysaccharides are very hydrophilic, those with lower molecular weights (e.g. dextrins) forming colloidal systems with water (section 10.5).

In the brewing of beer, the starch in cereal grains is hydrolysed to maltose by enzymes present in germinating grain (malt). The brew is then inoculated with yeast which transforms maltose to ethanol by fermentation:

$$C_6H_{12}O_6 \rightarrow 2C_2H_5OH + 2CO_2\uparrow.$$

As pointed out in section 13.1, care must be exercised to avoid the contamination of the fermentation mixture with micro-organisms causing the oxidation of ethanol to ethanoic acid (acetic acid). Spirits can be obtained by distillation of dilute solutions of alcohol (beer, wine, etc.).

13.6 CARBOXYLIC ACIDS AND LIPIDS

Carboxylic acid molecules contain a carbonyl group and a hydroxyl group attached to the same carbon atom at the end of a hydrocarbon chain, e.g.

$$
\begin{array}{c}
\text{O} \\
\parallel \\
\text{CH}_3\text{—C—OH} \\
\text{Ethanoic acid} \\
\text{(acetic acid).}
\end{array}
$$

Using the systematic method of naming, we delete the final -e in the name of the corresponding alkane and substitute -oic acid.

$$
\begin{array}{cccc}
& & & \text{O} \\
& & & \parallel \\
\text{Thus} & \text{CH}_4 & \text{gives} & \text{H—C—OH} \\
& \text{Methane} & & \text{Methanoic acid} \\
& & & \text{(formic acid)}
\end{array}
$$

$$
\begin{array}{cc}
\text{CH}_3 & \text{CH}_3 \\
| & | \\
\text{CH}_3\text{—CH—CH}_2\text{—CH}_2\text{—CH}_3 \text{ gives } & \text{CH}_3\text{—CH—CH}_2\text{CH}_2\text{CO}_2\text{H} \\
\text{2-methyl-pentane} & \text{4-methyl-pentanoic acid}
\end{array}
$$

where the carboxylic-acid-group-carbon is always numbered C1. Note also that non-systematic names are common, especially in naming acid derivatives like esters.

The carbon atom next to the carboxylic acid group may be referred to as the α carbon and the next, the β carbon:

$$\overset{\beta}{CH_3}-CH_2-\overset{\alpha}{CH_2}-CO_2H$$

Thus it is customary to call acids of the following general structure:

$$\overset{OH}{\underset{|}{R-CH}}-CH_2-CO_2H$$

β-hydroxy acids. In addition, the term *acyl* group is used to indicate

$$\overset{O}{\underset{\|}{R-C-}}$$

where R indicates an alkyl group.

TABLE 13.3. Properties of some carboxylic acids.

Name	Formula	Molecular weight	m.p. °C	b.p. °C	Water solubility (g 100 ml⁻¹)	Name of acyl group
Methanoic (Formic) acid	HCO_2H	46	8	101	*	Formyl
Ethanoic (Acetic) acid	CH_3CO_2H	60	17	118	*	Acetyl
Propanoic (Propionic) acid	$C_2H_5CO_2H$	74	−21	141	*	Propionyl
Butanoic (Butyric) acid	$C_3H_7CO_2H$	88	−4	164	*	–
Pentanoic (Valeric) acid	$C_4H_9CO_2H$	102	−34	186	4·97	–
Hexanoic acid (Caproic)	$C_5H_{11}CO_2H$	116	−2	205	1·08	–
Decanoic acid (Capric)	$C_9H_{19}CO_2H$	172	32	270	0·015	–
Hexadecanoic (Palmitic) acid	$C_{15}H_{31}CO_2H$	256	63	390	0·0007	–
Octadecanoic (Stearic) acid	$C_{17}H_{35}CO_2H$	284	72	360	0·0003	–

* Soluble in all proportions.

Properties and Reactions of Carboxylic Acids

Since hydrogen bonds can form between two carboxylic acid groups and between carboxylic acid molecules and water, the lower molecular weight acids have relatively high melting and boiling points and high water solubilities (those up to decanoic are liquids, Table 13.3). However, as in other homologous series the hydrophobic nature of the hydrocarbon chain increasingly dominates the physical properties of the acids as the molecular weight rises. Several of the lower acids have strong odours, e.g. Butanoic smells like rancid butter, whereas C_6 to C_{10} acids are said to smell like goats.

1. ACID/BASE PROPERTIES

The highly electronegative oxygen atoms in the hydroxyl and carbonyl functional groups attract electrons strongly away from the hydroxyl hydrogen atom, i.e.

$$
\begin{array}{cc}
O & e \\
\| & \leftarrow \\
R-C-O-H
\end{array}
$$

so that the hydrogen atom is not tightly held and tends to ionize in water:

$$
\begin{array}{cc}
O & O \\
\| & \| \\
R-C-O-H+H_2O \rightarrow R-C-O^-+H_3O^+
\end{array}
$$

giving *carboxylate ions* (formate, acetate, butyrate, stearate, etc.) whose chemical stability in solution also favours the ionization. Consequently, carboxylic acids are weak acids whose degree of dissociation varies according to the following K_a values:

Methanoic acid	$1\cdot 8 \times 10^{-4}$	
Ethanoic acid	$1\cdot 8 \times 10^{-5}$	(Compare with
Octanoic acid	$1\cdot 3 \times 10^{-5}$	Table 8.2.)

The neutralization of carboxylic acids by inorganic bases gives rise to a range of salts, many of which are useful; for example, sodium palmitate and sodium stearate are soaps, whereas lead acetate (sugar of lead) is a rather hazardous pain-killer.

2. ESTERIFICATION AND HYDROLYSIS OF ESTERS

As we noted in 13.1, alcohols react with carboxylic acids to give esters which have the general formula:

$$R^1-\overset{\overset{\textstyle O}{\|}}{C}-OR$$

For example
$$C_2H_5OH + CH_3CO_2H \rightarrow CH_3\overset{\overset{\textstyle O}{\|}}{C}-OC_2H_5 + H_2O$$
Ethyl acetate

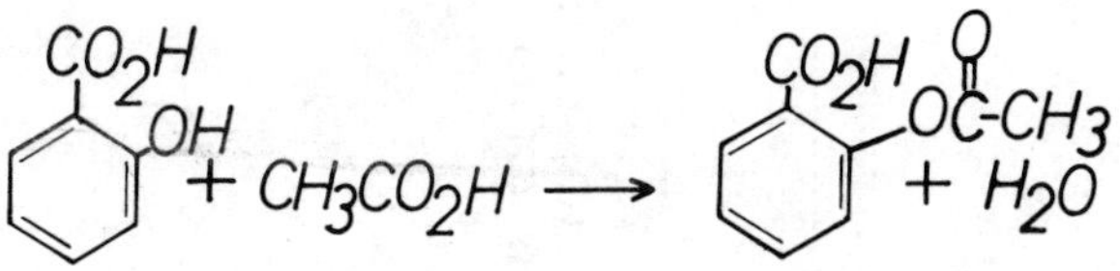

Salicylic acid

Acetyl salicylic (Aspirin) acid

Esters are important as flavouring materials, solvents, drugs, perfumes and as synthetic textiles (e.g. polyester fibres). Many have very pleasant smells which help in their identification. They can be *hydrolysed* back to the original acid and alcohol under alkaline conditions. In general:

$$\text{Alcohol} + \text{acid} \underset{\underset{\text{high pH}}{\text{Hydrolysis}}}{\overset{\overset{\text{low pH}}{\text{Esterification}}}{\rightleftharpoons}} \text{Ester} + H_2O$$

(compare with the hydrolysis of polysaccharides, section 13.5).

In biochemistry there are two groups of highly important complex esters—the triglyceride lipids and 'high energy' esters.

(a) *The Triglyceride Lipids*

Lipids are defined as substances, originating from living organisms, which are soluble in hydrophobic solvents (e.g. ether, chloroform, benzene) but insoluble in water. Lipids are important as energy stores in seeds and in the fat deposits of animals, but they have many other biochemical roles to play, for example, in determining the hydrophobic properties of cell membranes and in the electrical insulation of nerves. In spite of this rather loose definition, the majority of lipids belong to the same family, the triglycerides (carboxylic acid esters of glycerol), whose generalized structure is:

TABLE 13.4. The most widely distributed carboxylic acids occurring in the triglycerides of plants and animals.

Acid	Number of Carbon atoms	Formula
Oleic (Unsaturated)	18	$CH_3-(CH_2)_7-CH=CH-(CH_2)_7-CO_2H$
Linoleic*	18	$CH_3-(CH_2)_4-CH=CH-CH_2-CH=CH-(CH_2)_7-CO_2H$
Linolenic*	18	$CH_3-CH_2-CH=CH-CH_2-CH=CH-CH_2-CH=CH-(CH_2)_7-CO_2H$
Lauric (Saturated)	12	$CH_3-(CH_2)_{10}-CO_2H$
Palmitic	16	$CH_3-(CH_2)_{14}-CO_2H$
Stearic	18	$CH_3-(CH_2)_{16}-CO_2H$

* Essential fatty acids which must be provided in the diet of vertebrates (section 12.2).

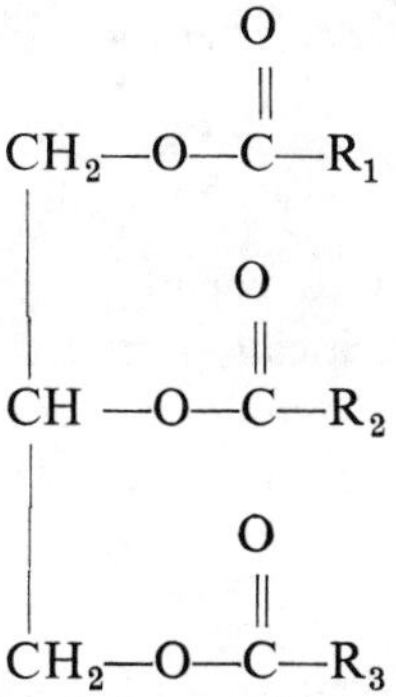

$$CH_2-O-\overset{\displaystyle O}{\overset{\|}{C}}-R_1$$

$$CH\ -O-\overset{\displaystyle O}{\overset{\|}{C}}-R_2$$

$$CH_2-O-\overset{\displaystyle O}{\overset{\|}{C}}-R_3$$

(R_1, R_2 and R_3 may be the same, or different groups)

where the R groups are straight alkane or alkene chains, usually of at least ten carbon atoms. Triglycerides which yield unsaturated or short-chain carboxylic ('fatty') acids on hydrolysis tend to be liquids at room temperature and are, therefore, called *oils* (e.g. many vegetable oils—sunflower, cottonseed, etc.), whereas those yielding saturated 'fatty acids' are solids at room temperature and are termed *fats* (e.g. various animal fats). (Note that care should be exercised to avoid confusion between the terms lipid, fat and oil; the term *fat*, in particular, is commonly employed in the place of lipid.)

Although the range of possible combinations of carboxylic acids with glycerol indicated by the general structure is very large, the majority of lipids from plants and animals contain combinations of only six carboxylic acids (Table 13.4). The 'hardening' of vegetable oils to fats is discussed in section 12.2 as are the techniques for determining the degree of unsaturation of an oil sample. The alkaline hydrolysis (or saponification) of saturated triglycerides yields glycerol and soaps (section 10.2):

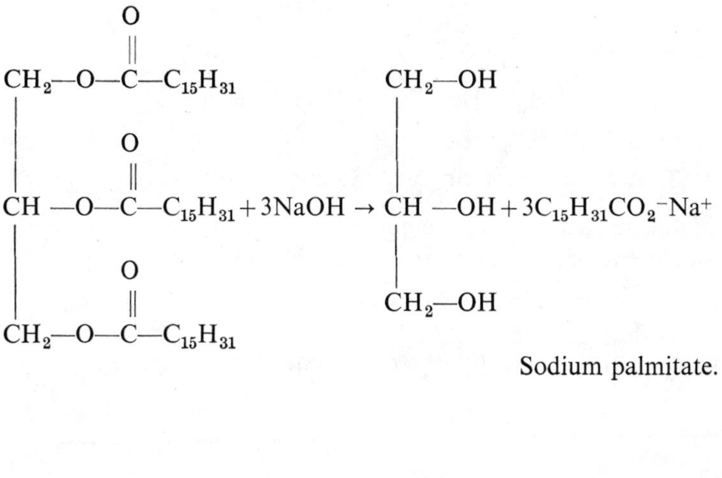

$$CH_2-O-\overset{\displaystyle O}{\overset{\|}{C}}-C_{15}H_{31}$$

$$CH\ -O-\overset{\displaystyle O}{\overset{\|}{C}}-C_{15}H_{31}+3NaOH \rightarrow$$

$$CH_2-O-\overset{\displaystyle O}{\overset{\|}{C}}-C_{15}H_{31}$$

$$CH_2-OH$$
$$CH\ -OH+3C_{15}H_{31}CO_2{}^-Na^+$$
$$CH_2-OH$$

Sodium palmitate.

Some more complex triglycerides are discussed in section 14.4.

(b) *High Energy Esters*

'High energy compounds', which can store chemical (potential) energy and release it when required, are critically important in the energy metabolism of living organisms. Examples include several esters, whose hydrolysis reactions are highly exergonic, such as:

Acyl phosphates

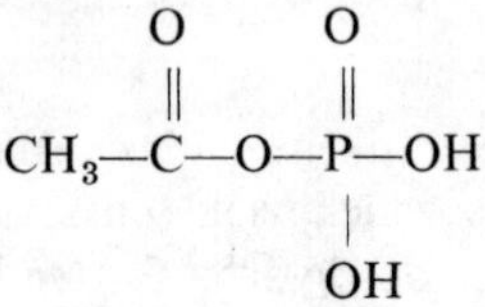

$$CH_3-\overset{\overset{\displaystyle O}{\|}}{C}-O-\overset{\overset{\displaystyle O}{\|}}{\underset{\underset{\displaystyle OH}{|}}{P}}-OH$$

(where Phosphoric acid is acting as an alcohol)

Acetyl phosphate

Thioesters

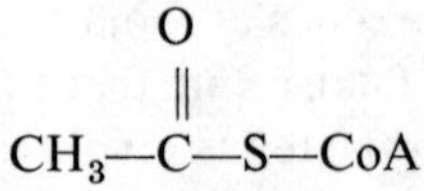

$$CH_3-\overset{\overset{\displaystyle O}{\|}}{C}-S-CoA$$

Acetyl Coenzyme A

In this molecule, the place of the alcohol is taken by a thiol whose structure is discussed in section 14.3.

3. DECARBOXYLATION REACTIONS

Simple carboxylic acids do not readily lose the carboxylic acid group in chemical reactions. However, the decarboxylation of β-keto acids is one step in the metabolism of lipids, e.g.

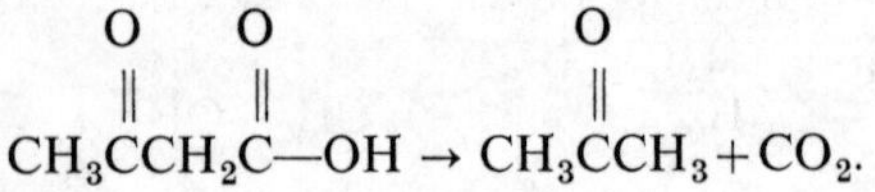

$$CH_3\overset{\overset{\displaystyle O}{\|}}{C}CH_2\overset{\overset{\displaystyle O}{\|}}{C}-OH \rightarrow CH_3\overset{\overset{\displaystyle O}{\|}}{C}CH_3 + CO_2.$$

4. THE REACTIONS OF METHANOIC ACID

Methanoic acid is unique amongst carboxylic acids since it contains an aldehyde as well as a carboxylic acid group:

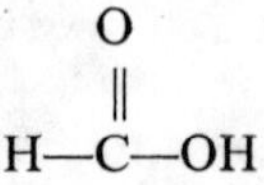

$$H-\overset{\overset{\displaystyle O}{\|}}{C}-OH$$

As a result, methanoic acid exhibits the reactions of both groups. For example, it is a weak acid but also gives positive Fehling's and Tollen's tests.

Important Carboxylic Acids, Their Occurrence and Use

Methanoic (Formic) Acid is responsible for the painfulness of ant bites and nettle stings. It is used for removing hair from hides, in dyeing and in making plastics.

Ethanoic (Acetic) Acid is used in dyeing, the manufacture of textiles and as a solvent. Vinegar is 4% ethanoic acid. The pure acid is called 'glacial acetic acid' since it freezes at 17°C.

Butanoic (Butyric) Acid is partly responsible for the taste of butter. It is added to margarine to make it more palatable.

Long-chain Acids are used in soaps, candles, floor wax, shoe polish, etc.

Other carboxylic acids of importance in biochemistry include:

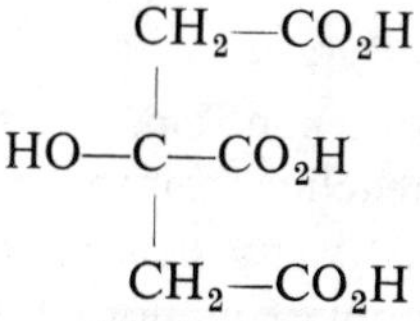

$$CH_2-CO_2H \qquad CH_2-CO_2H \qquad \qquad O$$
$$| \qquad\qquad\qquad | \qquad\qquad\qquad\qquad ||$$
$$HO-C-CO_2H \qquad CH_2-CO_2H \qquad CH_3-C-CO_2H$$
$$|$$
$$CH_2-CO_2H$$

Citric acid 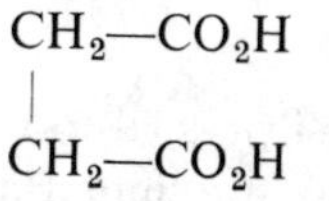　　　　　Succinic acid 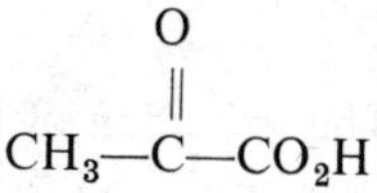　　　　Pyruvic acid

and

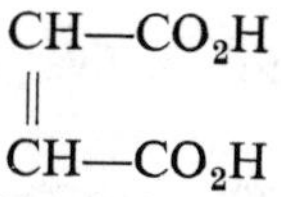

$$CH-CO_2H$$
$$||$$
$$CH-CO_2H$$

Fumaric acid

which take part in the Krebs Cycle;

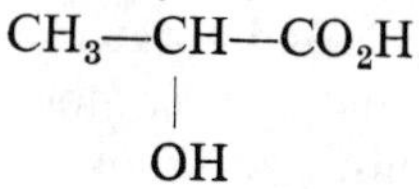

$$CH_3-CH-CO_2H$$
$$|$$
$$OH$$

Lactic acid

which accumulates in seeds and in muscles during anaerobic metabolism;

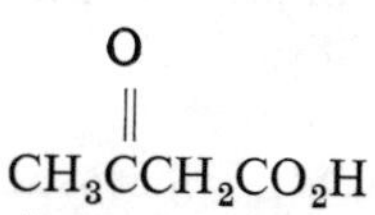

$$\overset{O}{\overset{||}{CH_3CCH_2CO_2H}} \qquad\qquad \overset{OH}{\overset{|}{CH_3-CH-CH_2-CO_2H}}$$

Acetoacetic acid 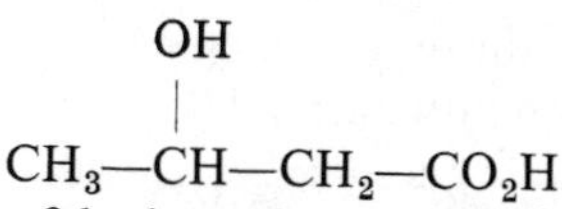　　　　　β-hydroxy-butyric acid which together with acetone, constitute the 'ketone bodies' which accumulate in the blood under conditions of rapid fat degradation (e.g. in acute form in diabetes);

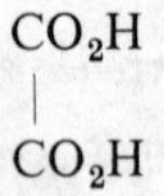

$$
\begin{array}{c}
CO_2H \\
| \\
CO_2H \\
\text{Oxalic acid}
\end{array}
\qquad
\begin{array}{c}
OH \\
| \\
CH\!-\!CO_2H \\
| \\
CH\!-\!CO_2H \\
| \\
OH \\
\text{Tartaric acid}
\end{array}
$$

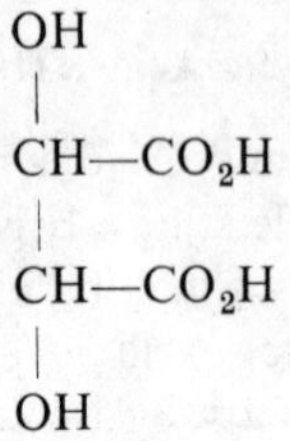

which occur at high concentrations in the leaves and fruits of certain plant species.

EXERCISES

(1) The oxidation of isocitric acid to oxalosuccinic acid, and malic acid to oxaloacetic acid are important reactions in the energy metabolism of cells. Consult a biochemistry textbook for the formulae of these four acids and write simple equations for the two reactions. Are these oxidations of primary, secondary or tertiary alcohols?

(2) Give systematic names for the five phenols at the end of section 13.2.

(3) Can hydrogen bonds occur between:
  (a) two aldehyde molecules;
  (b) two ketone molecules;
  (c) an aldehyde molecule and a water molecule;
  (d) a ketone molecule and a water molecule?

(4) Consult organic chemistry textbooks to find out what Fehling's and Tollen's reagents are; write equations for their reactions with aldehydes and ketones. What are the visual results of these reactions?

(5) Disaccharides, like monosaccharides, can be classed as reducing or non-reducing sugars. Why is maltose reducing whereas sucrose is non-reducing?

(6) Calculate the number of glucose molecules condensed together to give a macromolecule of amylopectin (MW 5 000 000)?.

FURTHER READING

The basic organic chemistry of oxygen containing functional groups is considered in the texts suggested for Chapter 11.

On carbohydrates:

1. PERCIVAL E.G.V. (1962) *Structural Carbohydrate Chemistry*, 2nd Edition. Garnet Miller.
2. PHELPS C.R. (1972) *Polysaccharides*. Oxford Biology Readers 27.

On lipids:

3. CHAPMAN D. (1969) *Introduction to Lipids*. McGraw Hill, New York.

For general information on nutrition (e.g. essential fatty acids) consult:

4. MCDONALD P., EDWARDS R.A. & GREENHALGH J.F.D. (1973) *Animal Nutrition*, 2nd Edition. Hafner.

For biochemical background:

5. DAVIES D.D., GIOVANELLI J. & AP REES T. (1964) *Plant Biochemistry*. Blackwell Scientific Publications, Oxford.
6. HARRISON K. (1965) *A Guide-Book to Biochemistry*, 2nd Edition. Cambridge University Press.

After the establishment of the fundamental ideas, the next step in the study of organic chemistry is to understand *how* reactions occur. A valuable introduction to this subject is given in:

7. SYKES P. (1965) *A Guidebook to Mechanism in Organic Chemistry*, 2nd Edition. Longman, London.

CHAPTER 14
ORGANIC COMPOUNDS
CONTAINING NITROGEN,
SULPHUR AND PHOSPHORUS

Of the three major groups of biochemical substances, *carbohydrates* (section 13.5) and *lipids* (13.6) are composed primarily of oxygen-containing organic molecules, but in order to understand the properties of *proteins*, it is also necessary to study the chemistry of nitrogen- and sulphur-containing functional groups. In this chapter we shall also consider some phosphorus-containing compounds which are important in heredity and in energy metabolism.

14.1 THE AMINES AND AMIDES

The amines are aliphatic and aromatic compounds containing one or more amino ($-NH_2$, $-NHR$, etc.) groups. They can be considered as derivatives of ammonia, with substitution of one, two or three hydrogen atoms by alkyl (or aryl) groups giving primary, secondary or tertiary amines:

NH_3	RNH_2	R_2NH	R_3N
Ammonia	Primary Amine	Secondary Amine	Tertiary Amine

e.g.	CH_3NH_2	$(CH_3)_2NH$	$(CH_3)_3N$
	Methylamine	Dimethylamine	Trimethylamine.

The bonding of alkyl groups to the amine nitrogen atom can go one step further to give *quaternary* ammonium ions (analogous to the ammonium ion, NH_4^+); for example, choline (a component of the complex phospholipid sphingomyelin from brain and muscles, section 14.4)—

$$CH_3-\overset{\overset{\displaystyle CH_3}{|}}{\underset{\underset{\displaystyle CH_3}{|}}{N^+}}-CH_2CH_2OH$$

180

As we can see from the examples above, simple amines are named like ethers, the attached groups being listed, followed by the name -amine. More complicated molecules can be named systematically, e.g.

$$\begin{array}{ccc} Cl & & NH_2 \\ | & & | \\ CH_3{-}CH{-}CH_2{-}CH{-}CH_2{-}OH \end{array}$$
2-amino-4-chloro-1-pentanol.

The simplest aromatic amine, amino-benzene, is normally called *Aniline*, but more complex aromatic amines are most conveniently named using the systematic method outlined in 12.4, for example—

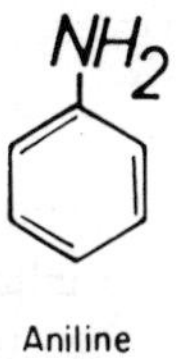

Aniline

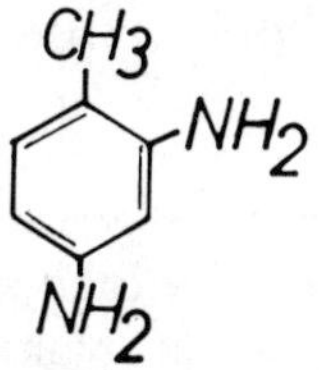

2,4–diamino
toluene

TABLE 14.1. Properties of some simple amines.

Name	Formula	Molecular weight	m.p. °C	b.p. °C	Water solubility (g 100 ml^{-1})	K_b
Ammonia	NH_3	17	-78	-33	89·9	$1·8 \times 10^{-5}$
Methylamine	CH_3NH_2	31	-94	-6	v. soluble	$4·4 \times 10^{-4}$
Ethylamine	$C_2H_5NH_2$	45	-81	17	*	$5·6 \times 10^{-4}$
Aniline	$C_6H_5NH_2$	93	-6	184	3·7	$3·8 \times 10^{-10}$

* Soluble in all proportions.

Properties and Reactions of Amines

In general, low molecular weight amines are strong-smelling gases (methylamine) or volatile liquids (ethylamine) whose high water solubilities are due to hydrogen bonding between the amino group and water.

1. ACID/BASE PROPERTIES

Like ammonia, amines are weak bases, for example, in the ionization of methylamine in water (giving a quaternary ammonium ion):

$$CH_3NH_2 + H_2O \rightleftharpoons CH_3NH_3^+ + OH^-,$$

$$K_b = \frac{[CH_3NH_3^+][OH^-]}{[CH_3NH_2]} = 4.4 \times 10^{-4}$$

showing that methylamine is a weaker base than sodium hydroxide ($K_b > 1$) but stronger than ammonia ($K_b = 1.8 \times 10^{-5}$) and aniline ($K_b = 3.8 \times 10^{-10}$) (see Exercise section).

2. THE DECOMPOSITION OF AMINES (KJELDAHL DETERMINATION OF ORGANIC NITROGEN)

When amines are digested with concentrated suphuric acid in the presence of a Cu/Se catalyst, the nitrogen atoms are liberated in the form of ammonium sulphate. If the solution is then treated with a strong base (NaOH), the ionization of the ammonium ion is suppressed and gaseous ammonia will be evolved when the solution is heated. This is the principle of the Kjeldahl determination of organic nitrogen in which the ammonia evolved is collected and measured by acid/base titration (see Chapter 8, Further Reading). The Kjeldahl method has wide applications in medicine, biochemistry, ecology and agriculture (for example in determining the nitrogen content of soil organic matter).

3. THE CONDENSATION OF AMINES WITH CARBOXYLIC ACIDS—THE AMIDES

Ammonia and amines react with carboxylic acids to give *amides*. In this type of reaction, the acid is normally in the form of an acid derivative (e.g. an ester) represented in generalized form as:

$$\begin{array}{c} O \\ \parallel \\ R\!-\!C\!-\!X \end{array}$$

For example:

$$\underset{\substack{\text{Ethanoic acid} \\ \text{derivative}}}{CH_3-\overset{\displaystyle O}{\overset{\|}{C}}-X} + NH_3 \rightarrow \underset{\text{Acetamide}}{CH_3-\overset{\displaystyle O}{\overset{\|}{C}}-NH_2} + HX$$

or more generally:

$$R-\overset{\displaystyle O}{\overset{\|}{C}}-X + R^1NH_2 \rightarrow \underset{\text{Amide}}{R-\overset{\displaystyle O}{\overset{\|}{C}}-NHR^1} + HX$$

Amides contain the *amide linkage*:

$$-\overset{\displaystyle O}{\overset{\|}{C}}-\overset{\displaystyle H}{\overset{|}{N}}- \quad \text{or} \quad -\overset{\displaystyle O}{\overset{\|}{C}}-\overset{\displaystyle R}{\overset{|}{N}}-$$

which is called the *peptide linkage* in peptides and proteins. Amides can be hydrolysed to give back acid and amine (or ammonia) under both acid and basic conditions.

Acid Hydrolysis

e.g.

$$R-\overset{\displaystyle O}{\overset{\|}{C}}-NH_2 + HCl + H_2O \rightarrow R-\overset{\displaystyle O}{\overset{\|}{C}}-OH + NH_4^+Cl^-.$$

Base Hydrolysis

e.g.

$$R-\overset{\displaystyle O}{\overset{\|}{C}}-N(CH_3)_2 + NaOH \rightarrow R-\overset{\displaystyle O}{\overset{\|}{C}}-O^-Na^+ + (CH_3)_2NH.$$

The most important simple amide in biochemistry is *Urea*:

$$\begin{array}{c} NH_2 \\ | \\ C = O \\ | \\ NH_2 \end{array}$$

which is the chief end-product of mammalian protein metabolism and a major constituent of urine. It is a useful nitrogen fertilizer substance, yielding ammonium ions in soil and may also be used as a concentrated nitrogen source in livestock feeds.

When heated, urea forms a condensation product, *biuret*

$$NH_2-\overset{\overset{\displaystyle O}{\|}}{C}-NH_2 + NH_2-\overset{\overset{\displaystyle O}{\|}}{C}-NH_2 \rightarrow$$

$$NH_2-\overset{\overset{\displaystyle O}{\|}}{C}-NH-\overset{\overset{\displaystyle O}{\|}}{C}-NH_2 + NH_3$$
Biuret

which gives a distinctive purple product with alkaline copper sulphate solution. (On heating, proteins also give a positive *Biuret Test*.) Biuret is not toxic to livestock but concentrations in fertilizer urea must be kept below 1% to avoid phytotoxicity.

Some Important Amines

(1) The α-amino acids which are discussed fully in section 14.2.
(2) Several simple aliphatic amines account for the strong smells of fish and fish products, e.g. CH_3NH_2, $(CH_3)_2NH$, $(CH_3)_3N$, $C_2H_5NH_2$, etc. Tertiary amines, which can occur in bad silage, are toxic to ruminants.
(3) We have already seen that choline occurs in muscle tissues. Another quaternary ammonium compound, *acetylcholine* (a derivative of choline), is a chemical transmitter of nerve impulses:

$$CH_3-\overset{\overset{\displaystyle CH_3}{|}}{\underset{\underset{\displaystyle CH_3}{|}}{N^+}}-CH_2CH_2-O-\overset{\overset{\displaystyle O}{\|}}{C}-CH_3.$$

(4) The synthetic diamine EDTA (ethylene diamine tetra-acetic acid) is a chelating agent widely used in the dissolution and chemical analysis of sparingly soluble cations—

(as an anion, chelating a Calcium cation)

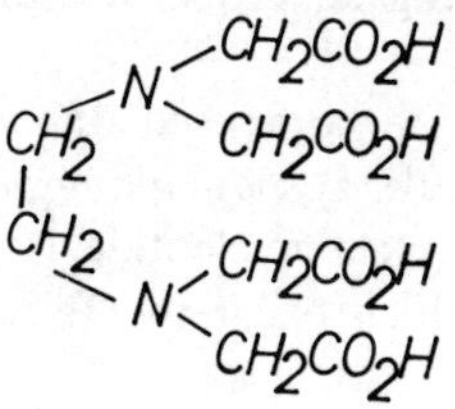

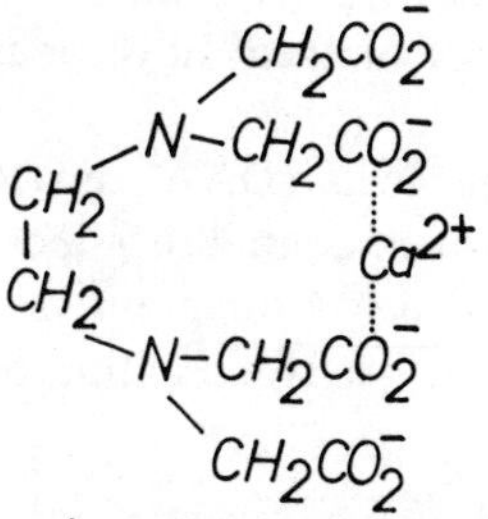

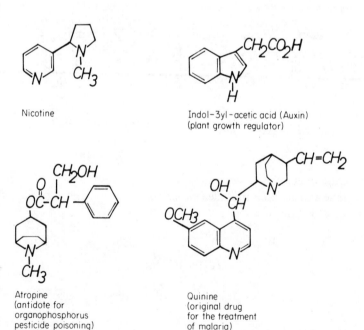

(5) There exists a great variety of aromatic compounds in which one or two carbon atoms of the benzene ring have been replaced by nitrogen atoms. These *heterocyclic* aromatic amines include a large number of important compounds, for example pyrimidine and purine (section 12.4):

Nicotine

Indol–3yl–acetic acid (Auxin)
(plant growth regulator)

Atropine
(antidote for organophosphorus pesticide poisoning)

Quinine
(original drug for the treatment of malaria)

Nicotine, atropine and quinine belong to the large group of naturally occurring amine bases called alkaloids, which can be obtained from the extracts of plant tissues (notably the Umbelliferae and Solan-aceae); most of them have pharmacological properties (drugs and poisons).

(6) Nucleic acids (DNA and RNA) are composed of *nucleotides* which, in turn, contain a *ribose molecule*, a *phosphate group* and a *complex amine*. A diagram of a complete nucleotide is given in section 14.4. The four amine bases which occur in DNA are—

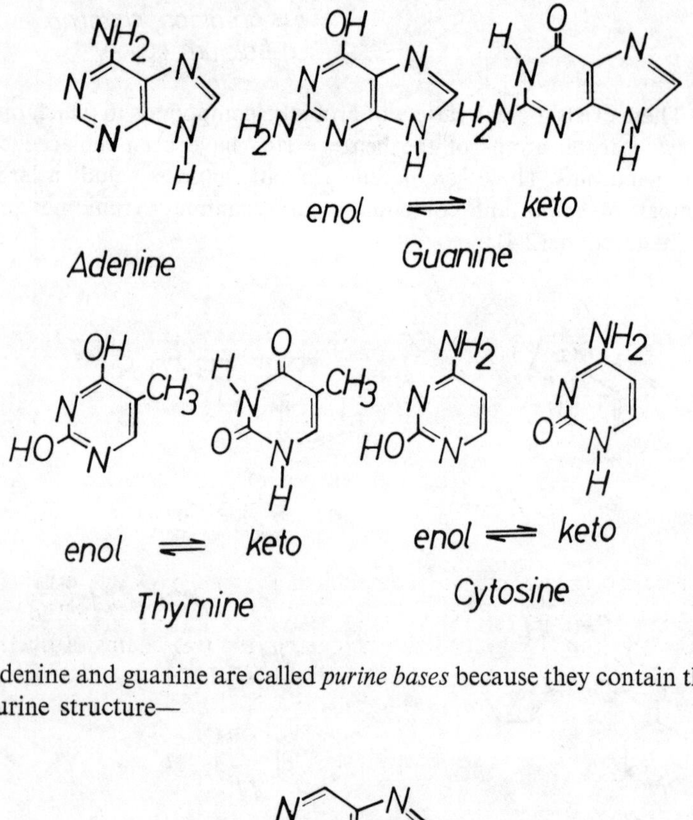

Adenine

enol ⇌ keto

Guanine

enol ⇌ keto

Thymine

enol ⇌ keto

Cytosine

Adenine and guanine are called *purine bases* because they contain the purine structure—

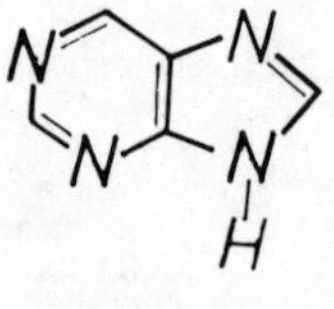

whereas thymine and cytosine are *pyrimidine bases*. When not attached to the other groups in a nucleotide, guanine, thymine and cytosine tend to take up the enol structure which is the more stable tautomer for these molecules (section 12.3). However, in DNA they take up the keto form since this allows adenine to form strong hydrogen bonds with thymine, and guanine with cytosine—

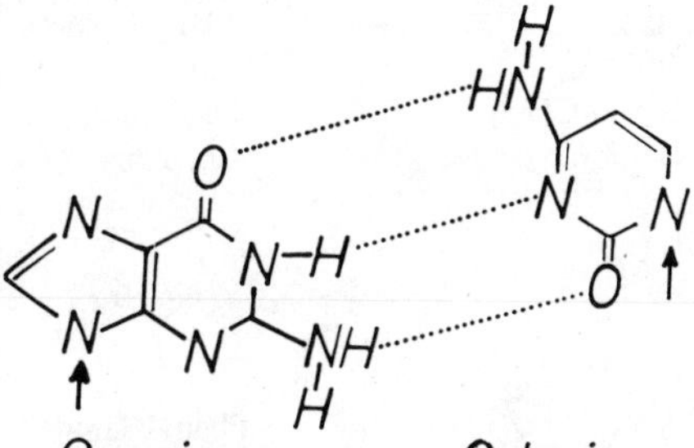

Guanine   Cytosine

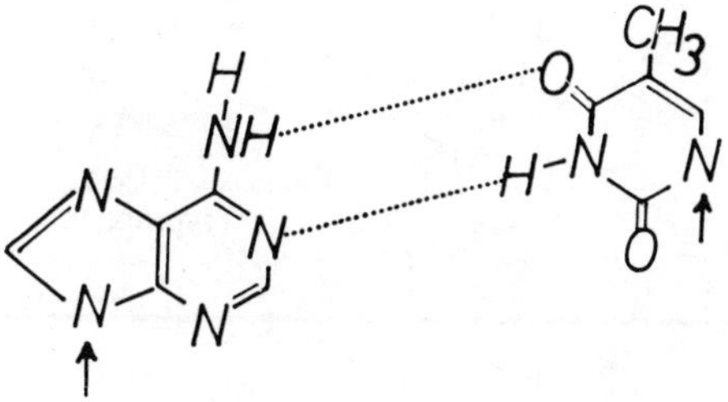

Adenine   Thymine

(where the arrows indicate the points of attachment of the remainder of each nucleotide molecule). These hydrogen bonds linking the base pairs are responsible for holding together the two chains of nucleotides in the *double helix* structure (see Figure 14.3).

14.2 AMINO ACIDS, PEPTIDES AND PROTEINS

Proteins and peptides are macromolecules formed by the condensation of a large number of simpler molecules called α-amino acids, whose general formula is:

$$\overset{\displaystyle R}{\underset{\displaystyle H_2N\!-\!CH\!-\!CO_2H}{|}}$$

TABLE 14.2. Classification of α-amino acids.

Number	R group	Name	3-letter Name
Class 1.	*R is hydrophobic*		
1	$-H$	Glycine	GLY
2	$-CH_3$	Alanine	ALA
3	$-CH(CH_3)_2$	Valine	VAL
4	$-CH_2CH(CH_3)_2$	Leucine	LEU
5	$-CHCH_2CH_3$ $\quad\mid$ CH_3	Isoleucine	ILE
6	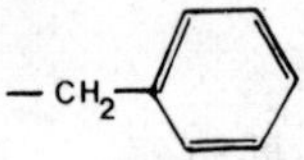	Phenylalanine	PHE
7	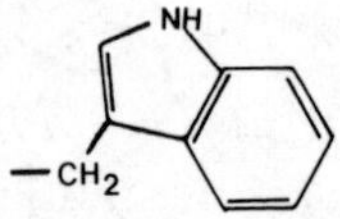	Tryptophan	TRP (or TRY)
Class 2.	*R is hydrophilic, containing an alcohol or phenol group*		
8	$-CH_2OH$	Serine	SER
9	$-CHOH$ $\quad\mid$ CH_3	Threonine	THR
10	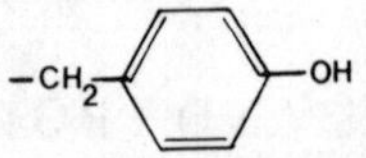	Tyrosine	TYR
Class 3.	*R is hydrophilic, containing a carboxylic acid or amide group*		
11	$-CH_2CO_2H$	Aspartic Acid	ASP
12	$-CH_2CH_2CO_2H$	Glutamic Acid	GLU
13	$-CH_2CONH_2$	Asparagine	ASN
14	$-CH_2CH_2CONH_2$	Glutamine	GLN

TABLE 14.2 (Cont.)

Number	R group	Name	3-letter Name
Class 4. *R is hydrophilic, containing an amine group**			
15	$-CH_2CH_2CH_2CH_2NH_2$	Lysine	LYS
16	$-CH_2CH_2CH_2NH-\overset{\overset{NH}{\|\|}}{C}-NH_2$	Arginine	ARG
17	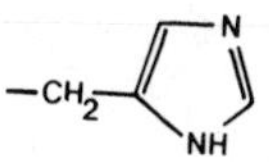	Histidine	HIS

* Note that although Tryptophan carries an amine group, its hydrophobic properties are determined by the attached benzene ring.

Class 5. *R contains sulphur†*			
18	$-CH_2SH$	Cysteine	CYS
19	$-CH_2-S-S-CH_2-$	Cystine	CYS[1] or CYS–CYS
20	$-CH_2CH_2S\ CH_3$	Methionine	MET

† Cysteine is hydrophilic and Methionine hydrophobic.

Class 6. *The imino acids‡* (whole molecule shown)			
21	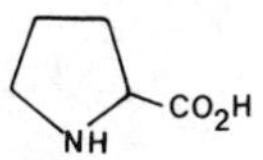	Proline	PRO
22	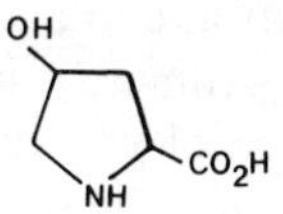	Hydroxyproline	HYP

‡ In which the amino group forms part of a ring structure.

where the R group is attached to the α carbon of the carboxylic acid. (Compare with carbohydrate macromolecules which are composed of monosaccharide units.)

In theory, there can exist an enormous number of possible amino acids (e.g. R alkyl or aryl, branched or unbranched, saturated or unsaturated). However, in proteins from living organisms, there are only 22 common R groups, giving the 22 corresponding α-amino acids shown in Table 14.2.

Properties and Reactions of Amino Acids

1. ACID/BASE PROPERTIES

Since amino acid molecules contain both acidic and basic groups, they behave as amphoteric substances, as previously shown in section 8.5.

2. TRANSAMINATION AND ESSENTIAL AMINO ACIDS

In their roots and leaves, photosynthetic plants reduce absorbed nitrate to ammonia, which is incorporated into the following amino acids: Aspartic Acid, Alanine, Glutamic Acid.

Other amino acids can then be synthesized from these acids by the process of *transamination*. In general terms:

$$\underset{\alpha\text{-keto acid}}{R-\overset{\overset{\displaystyle O}{\|}}{C}-CO_2H} + \underset{\alpha\text{-Amino acid}}{R^1-\overset{\overset{\displaystyle NH_2}{|}}{C}H-CO_2H} \rightarrow$$

$$R-\overset{\overset{\displaystyle NH_2}{|}}{C}H-CO_2H + R^1-\overset{\overset{\displaystyle O}{\|}}{C}-CO_2H$$
new α-amino acid.

Plants can synthesize the entire range of amino acids but animals, in general, cannot and are dependent ultimately upon plants for certain *essential amino acids*. The range of essential amino acids varies amongst species; for example, the pig requires: arginine, histidine, isoleucine, leucine, lysine, methionine, phenylalanine, threonine, tryptophan and valine, whereas poultry require all of these and glycine in addition.

3. FORMATION OF AMIDE OR PEPTIDE LINKS

In the last section, we noted that amines react with carboxylic acids to give amides, i.e.

$$RCO_2H + R^1NH_2 \rightarrow RCONHR^1.$$

Since they contain both amine and carboxylic acid groups, two amino acid molecules can condense together by this reaction to give what is called a dipeptide, e.g.

$$\begin{array}{c} CH_3 \\ | \\ H_2N{-}CH_2{-}CO_2H + H_2N{-}CH{-}CO_2H \rightarrow \\ \quad GLY \qquad\qquad\qquad ALA \end{array}$$

$$\begin{array}{c} O \quad H \;\; CH_3 \\ \| \quad | \quad\; | \\ H_2N{-}CH_2{-}C{-}N{-}CH{-}CO_2H \\ \text{GLY-ALA (dipeptide)} \end{array}$$

where the amide link between the two amino acids is called a peptide link. A dipeptide is still an amino acid (although not an α-amino acid) and it can condense with a further α-amino acid to give a tripeptide, e.g. GLY-ALA-LEU. Obviously this can go on to give very long chains of amino acids; peptides containing less that 100 α-amino acids are normally called polypeptides, whereas those with 100 to 10 000 or more are proteins. The peptide link in peptides and proteins may be broken to give back the original α-amino acids by acid hydrolysis or by the action of enzymes.

Peptides and proteins are of enormous importance in biochemistry: as hormones controlling metabolism and physiology; as enzymes regulating metabolic reactions; as respiratory pigments transporting oxygen to sites of respiration; in the immune response and as structural elements in muscles, joints and connective tissues.

The Structure of Peptides and Proteins

The *primary structure* of peptides and proteins is simply the *order* in which the α-amino acids are assembled; for example, the peptide hormone, *Oxytocin* (released from the pituitary—causes birth contractions in the uterus) has the following primary structure:

$$H_2N{-}GLY{-}LEU{-}PRO{-}CYS{-}ASN{-}GLN{-}ILE{-}TYR{-}$$
$$CYS{-}CO_2H$$

where the ends carrying *free* amine and carboxylic acid groups are indicated by —NH_2 and —CO_2H respectively.

Over the last thirty years, the primary structures of a large number of polypeptides and smaller proteins have been elucidated; for example the hormone insulin (51 amino acids); the enzyme ribonuclease (124) and the respiratory pigment myoglobin (153) (Figure 14.2). Some idea of the variation in amino acid composition amongst proteins is given by Table 14.3.

TABLE 14.3. The distribution of nine amino acids in selected proteins (g 100 g^{-1} of protein). (From *Chemical Background for the Biological Sciences*, 2nd Edition, by E.H. White (1970). Prentice-Hall, New Jersey.)*

Protein	GLY	ALA	VAL	LEU	ILE	MET	PHE	TRP	LYS
Fibroin (silk)	44	30	4	1	1	0	3	0	1
Keratin (wool)	7	4	5	11	0	1	4	2	3
Albumin (hen)	3	7	7	9	7	5	8	1	6
Haemoglobin (horse)	6	7	9	15	0	1	8	2	9
Insulin (ox)	4	5	8	13	3	0	8	0	3

* Adapted from L.F. Fieser & M. Fieser, *Advanced Organic Chemistry* (1961). Reinhold, New York.

A chain of linked amino acids does not normally remain in extended form but tends to coil up into an α-helix (Figure 14.1) which, by maximizing the number of intramolecular hydrogen bonds formed, is its most stable configuration. However, this coiling is not complete in most proteins and the *secondary structure* of the macromolecule is a measure of the extent and position of helical and extended chain sections. For example, only 70 % of the chain length of myoglobin is helical (Figure 14.2). The other important type of secondary structure is the pleated sheet which occurs in fibrous proteins like silk and keratin (hair and nails).

In most proteins, the helical chain folds up into an apparently irregular, but exactly defined, three-dimensional shape which is essential for the function of the protein (as an enzyme, respiratory pigment or in structural tissues like muscles and tendons). This irregular folding of the helical chain is the *tertiary structure* of the

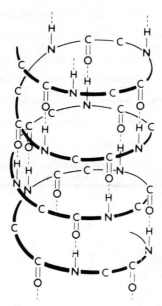

FIGURE 14.1. The α-helical structure of peptide and protein chains, held together by intramolecular hydrogen bonds. (From Dearden J.C. (1968) The hydrogen bond. *New Scientist*, **37**, 629.)

protein and is caused by interactions between functional groups attached to different parts of the chain (Table 14.2) i.e.

(a) attraction between hydrophobic groups;

(b) hydrogen bonding between hydrophilic groups, and

(c) disulphide bonds between cysteine molecules (section 14.3).

For example, Figure 14.2 illustrates the three-dimensional tertiary structure of myoglobin, a respiratory pigment involved in the storage of oxygen in tissues (especially in diving organisms like whales). In common with a number of respiratory pigments and enzymes, myoglobin contains a non-protein component, in this case an oxygen-binding haem group, embedded in the tertiary structure. Such inclusions are termed prosthetic groups.

In certain protein molecules, which are made up of a number of separate helical chains bound together by functional group interactions rather than peptide linkages, the three dimensional arrangements of the chains make up the *quaternary* structure of the molecule. For example, each haemoglobin molecule is composed of two 'α chains' and two 'β chains', each chain containing a haem prosthetic group embedded in its tertiary structure.

H-Val-Leu-Ser-Glu-Gly-Glu-Trp-Gln-Leu-Val-Leu-His-Val-Tyr-Ala-Lys-Val-
1   10
Glu-Ala-Asp-Val-Ala-Gly-His-Gly-Gln-Asp-Ile-Leu-Ile-Arg-Leu-Phe-Lys-
    20   30
Ser-His-Pro-Glu-Thr-Leu-Glu-Lys-Phe-Asp-Arg-Phe-Lys-His-Leu-Lys-Thr-
        40   50
Glu-Ala-Glu-Met-Lys-Ala-Ser-Glu-Asp-Leu-Lys-Gly-His-His-Glu-Ala-Glu-
                            60
Leu-Thr-Ala-Leu-Gly-Ala-Ile-Leu-Lys-Lys-Lys-Gly-His-His-Glu-Ala-Glu-
    70   80
Leu-Lys-Pro-Leu-Ala-Gln-Ser-His-Ala-Thr-Lys-His-Lys-Ile-Pro-Ile-Lys-Tyr-
        90   100
Leu-Glu-Phe-Ile-Ser-Glu-Ala-Ile-Ile-His-Val-Leu-His-Ser-Arg-His-Pro-Gly-.
        110   120
Asn-Phe-Gly-Ala-Asp-Ala-Gln-Gly-Ala-Met-Asn-Lys-Ala-Leu-Glu-Leu-Phe-
                            130
Arg·Lys-Asp-Ile-Ala-Ala-Lys-Tyr-Lys-Glu-Leu-Gly-Tyr-Gln-Gly-OH
    140   150   153

(a)

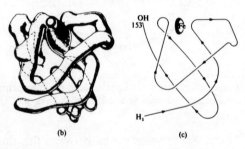

(b)   (c)

FIGURE 14.2. The primary, secondary and tertiary structure of whale myoglobin. (Adapted from J.C. Kendrew *et al.* (1960). Structure of Myoglobin. *Nature* **185**, 422. ©1960 Macmillan Journals Ltd.).

Factors such as heat, radiation or light, which cause changes in the secondary, tertiary or quaternary structures of a protein, without breaking peptide linkages between amino acids, are said to cause *denaturation* of the protein. A familiar example is the denaturation of egg white on boiling

14.3 THE THIOLS

Thiols are organic compounds containing one or more thiol (SH) groups and are sometimes called thioalcohols or mercaptans. In general, thiols are of little importance in elementary chemistry; however, because of the existence of an important thiol amino acid, cysteine:

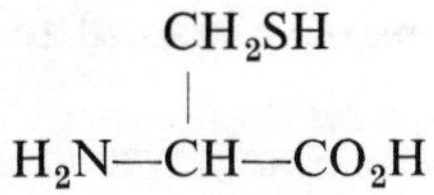

$$CH_2SH$$
$$|$$
$$H_2N—CH—CO_2H$$

it is essential to have some knowledge of the chemistry of the thiol functional group.

Properties and Reactions of Thiols

1. ACID/BASE PROPERTIES

Thiols are very weak acids, but slightly stronger than alcohols, e.g.

$$C_2H_5SH + H_2O \rightleftharpoons C_2H_5S^- + H_3O^+ \qquad K_a = 10^{-12}.$$

Under suitable conditions, salts of thiols do occur. This property is used in cell biology, where proteins containing cysteine may be precipitated, and, therefore, inactivated, as insoluble lead salts.

2. OXIDATION TO FORM DISULPHIDES

Under mild oxidizing conditions, two thiol molecules may react together to give a disulphide:

$$R—SH + HS—R \rightarrow R—S—S—R.$$

As noted in section 14.2, such disulphide bonds are involved in determining the tertiary structures of many proteins (e.g. insulin, and keratin in hair) by bonding together different regions of the chain of amino acids. In a simpler peptide, oxytocin, (section 14.2), the formation of a disulphide bond changes the molecule from the inactive, straight-chain form, to the active, bent-chain form:

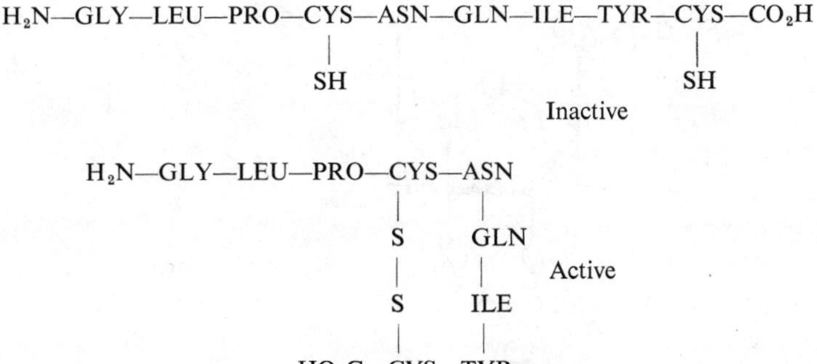

```
H₂N—GLY—LEU—PRO—CYS—ASN—GLN—ILE—TYR—CYS—CO₂H
                    |                        |
                    SH                       SH
                                        Inactive
```

```
    H₂N—GLY—LEU—PRO—CYS—ASN
                      |     |
                      S    GLN
                      |     |      Active
                      S    ILE
                      |     |
             HO₂C—CYS—TYR
```

Here, two cysteine groups have reacted to give one cystine (Table 14.2).

The disulphide bond is also employed in the biochemical oxidizing agent, Lipoic acid:

$$CH_2—CH_2—CHCH_2CH_2CH_2CH_2CO_2H \rightleftharpoons$$

| |
S————————S

Oxidized

$$CH_2—CH_2—CHCH_2CH_2CH_2CH_2CO_2H$$

| |
SH   SH

Reduced

3. ESTERIFICATION

As noted in section 13.6, thiols can substitute for alcohols in esterification reactions, i.e.

$$R—\overset{\overset{\displaystyle O}{\|}}{C}—OH + R^1—SH \rightarrow R—\overset{\overset{\displaystyle O}{\|}}{C}—SR^1 + H_2O.$$

In particular, the complex thiol, Co-enzyme A—

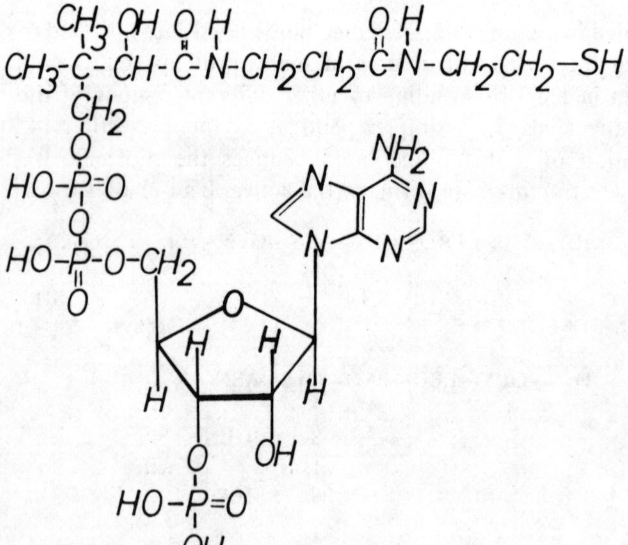

gives the thioester acetyl Co-enzyme A which participates widely in energy metabolism and biochemical syntheses (section 13.6).

Other thiols of biological importance include the foul-smelling butanethiol, C_4H_9SH, exuded by the skunk to repel predators. Note that the α-amino acid methionine is a thioether (Table 14.2).

14.4 ORGANIC COMPOUNDS CONTAINING PHOSPHORUS

Many compounds involved in biochemical reactions (energy and lipid metabolism) and in heredity are derivatives of phosphoric acid or one of its condensed forms (pyrophosphoric acid and triphosphoric acid):

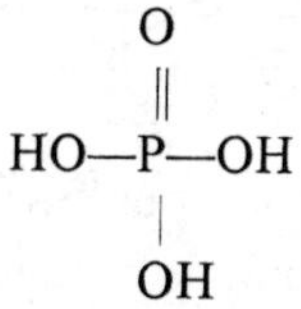

$$
\begin{array}{cc}
\underset{\text{phosphoric acid}}{\text{HO}-\overset{\displaystyle O}{\overset{\|}{\underset{\underset{\text{OH}}{|}}{P}}}-\text{OH}} &
\underset{\text{pyro- or diphosphoric acid}}{\text{HO}-\overset{\displaystyle O}{\overset{\|}{\underset{\underset{\text{OH}}{|}}{P}}}-O-\overset{\displaystyle O}{\overset{\|}{\underset{\underset{\text{OH}}{|}}{P}}}-\text{OH}}
\end{array}
$$

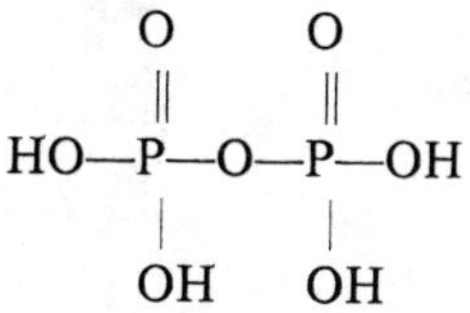

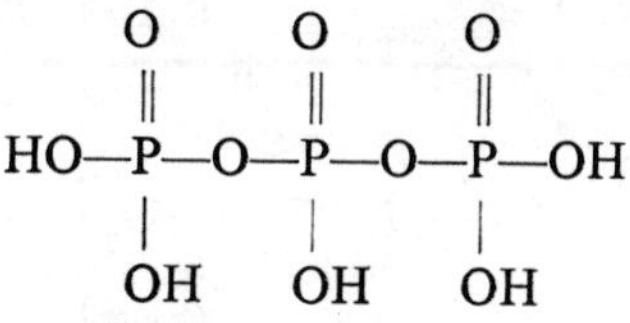

$$
\underset{\text{triphosphoric acid}}{\text{HO}-\overset{\displaystyle O}{\overset{\|}{\underset{\underset{\text{OH}}{|}}{P}}}-O-\overset{\displaystyle O}{\overset{\|}{\underset{\underset{\text{OH}}{|}}{P}}}-O-\overset{\displaystyle O}{\overset{\|}{\underset{\underset{\text{OH}}{|}}{P}}}-\text{OH}}
$$

where the hydroxyl groups of the acids can act as alcohol groups. For example, all of the carbohydrate compounds involved in the Embden-Meyerhof-Parnas Pathway (glycolysis or fermentation) are in the form of phosphates and the first step of the pathway is the phosphorylation of glucose to glucose-6-phosphate:

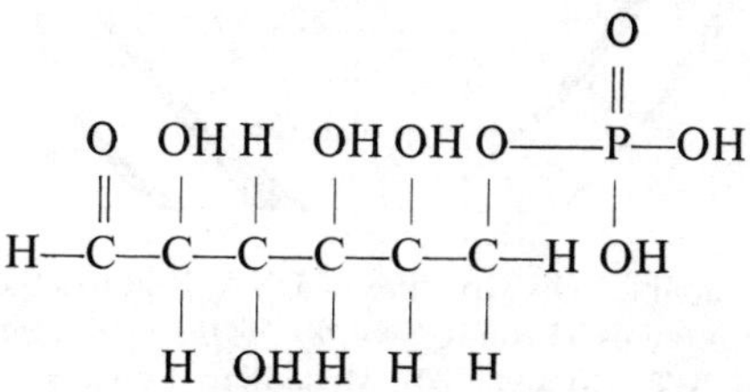

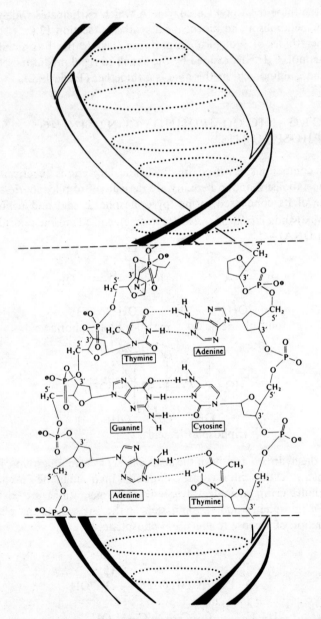

FIGURE 14.3. The double helix structure of DNA, held together by inter-molecular hydrogen bonds. (From *Chemistry for the Life Sciences*, 2nd edition, by J.G. Dawber & A.T. Moore (1980). Macmillan, London.)

catalysed by the enzyme hexokinase. (This reaction is analogous to the condensation of two alcohols, sections 13.1 and 13.5.)

In general, the chemistry of phosphorylated organic compounds is rather complex. However, even at an elementary level, it is essential to be able to recognize the following groups of phosphorus-containing biochemical substances.

(a) Nucleic Acids

As noted in section 14.1, nucleic acids are polymers of four different nucleotides. In DNA, these are *adenosine phosphate* (contains adenine); *thymidine phosphate* (thymine); *guanosine phosphate* (guanine) and *cytidine phosphate* (cytosine), e.g. adenosine phosphate (from DNA)—

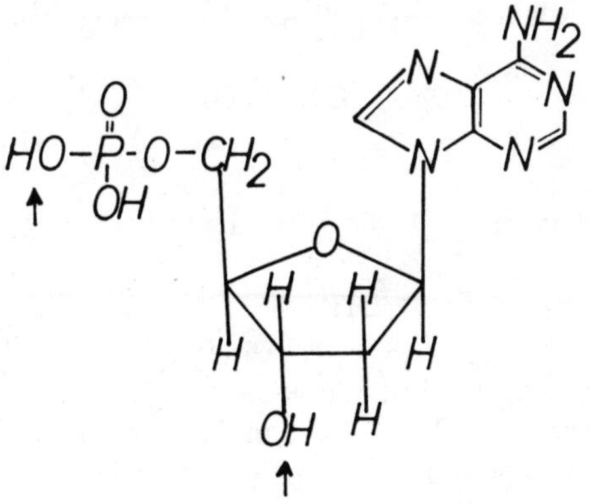

where the arrows indicate the points of attachment to adjacent nucleotides giving a single chain which pairs with a complementary single chain to give the hydrogen-bonded double helix of DNA (Figure 14.3).

(b) High Energy Compounds

Replacing the single phosphate group in adenosine phosphate (adenosine monophosphate or AMP) with a pyrophosphate or triphosphate group, we obtain adenosine diphosphate (ADP) or adenosine triphosphate (ATP). Because the hydrolysis reactions:

(1)

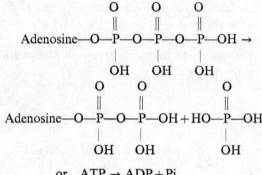

$$\text{or} \quad ATP \rightarrow ADP + Pi$$

(2)

or $ATP \rightarrow AMP + PPi$

are highly exergonic ($2 \cdot 9 \times 10^4$ J mole^{-1} for equation 1), ATP is a 'high energy compound' (section 13.6), used in metabolism to store and transfer energy. It is conventional to write ATP as:

where $\sim$ indicates a 'high energy bond' whose hydrolysis releases a large amount of energy.

(c) Phospholipids

In addition to the simple triglycerides discussed in section 13.6, there also exist in membranes, seeds, and, particularly in brain and nerves, a variety of more complex triglycerides (phospholipids) varying in structure from the relatively simple phosphatides:

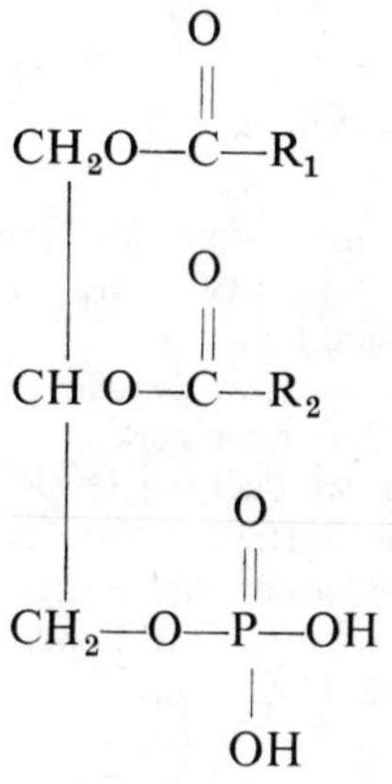

$$
\begin{array}{c}
O \\
\| \\
CH_2O-C-R_1 \\
| \\
O \\
\| \\
CH\ O-C-R_2 \\
| \\
O \\
\| \\
CH_2-O-P-OH \\
| \\
OH
\end{array}
$$

to sphingomyelins, e.g.:

$$
\begin{array}{c}
OH \\
| \\
CH-CH=CH-(CH_2)_{12}-CH_3 \\
| \\
O \\
\| \\
CH-NH-C-(CH_2)_{16}-CH_3 \\
| \\
O \\
\| \\
CH_2-O-P-O-CH_2-CH_2-N^+(CH_3)_3 \\
| \\
O_-
\end{array}
$$

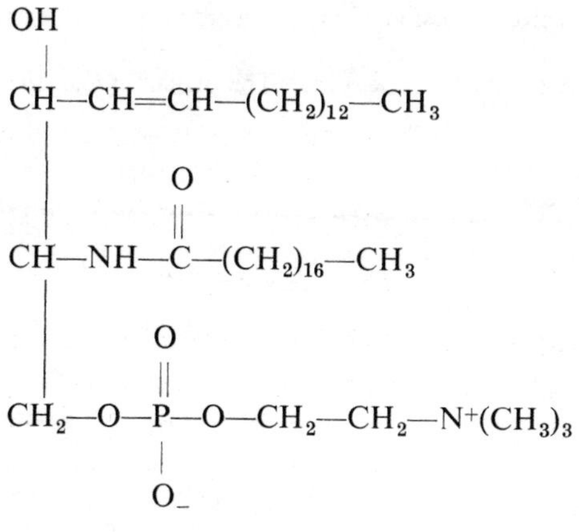

EXERCISES

(1) If 'aromatic alcohols' (phenols) are much stronger acids than aliphatic alcohols, why are aromatic amines much weaker bases than aliphatic amines?

(2) Several of the α-amino acids can be identified by the distinctive colours they give when treated with test reagents. Using textbooks, discover which α-amino acids give positive colour reactions in each of the following tests: (a) Ninhydrin test; (b) Biuret test; (c) Test with

alcoholic α-naphthol; (d) Purple ring test; (e) Lead acetate test; (f) Xanthoproteic test.

FURTHER READING

On the chemistry and biochemistry of peptides, proteins and nucleic acids:
1. DAWBER J.G. & MOORE A.T. (1980) *Chemistry for the Life Sciences*, 2nd Edition, Chs 7–9. Macmillan, London.
2. HOLUM J.R. (1969) *Introduction to Organic and Biological Chemistry*, Chs 10 and 17. John Wiley & Son, New York.
3. WHITE E.H. (1970) *Chemical Background for the Biological Sciences*, 2nd Edition, Chs 5 and 6. Prentice-Hall, Inc, New Jersey.
4. WYNN C.H. (1979) *The Structure and Function of Enzymes*, 2nd Edition. Studies in Biology 42. Edward Arnold, London.
For an account of the use of EDTA in chemical analysis, consult (for example):
5. GOLTERMAN H.L., CLYMO R.S. & OHNSTAD M.A.M. (1978) *Methods for Physical and Chemical Analysis of Fresh Waters*, 2nd Edition. IBP Handbook No. 8. Blackwell Scientific Publications, Oxford.
For a comprehensive discussion of energy metabolism ('high energy bonds', etc.) read:
6. LEHNINGER A.L. (1965) *Bioenergetics*. The Benjamin Co., Inc., New York.

PART 3
AGRICULTURAL CHEMICALS

CHAPTER 15
FERTILIZERS

Crop plants (like natural plant species) require adequate supplies of between 16 and 23 chemical elements for healthy growth and high yield. These elements can be classified into four groups:

(1) H, C, O—supplied by the atmosphere and water;

(2) N, Mg, P, S, K, Ca—the *Macronutrients* (some authorities include Fe in this group). Required in large quantities (i.e. at kg ha^{-1} levels in soils);

(3) B, Cl, Mn, Fe, Cu, Zn, Mo—the *Micronutrients* or *Trace Elements* required in small quantities (at ppm levels in soils);

(4) Na, Si, V, Co, Ni, Se, W—elements involved in the growth or metabolism of certain species, e.g. Legumes require Co; Na appears to be essential for plants using the C_4 pathway in photosynthesis.

Although several of these essential elements are abundant overall in the earth's crust (Mg 2·1%, K 2·6%, Ca 3·6%, Fe 5·1%), any of the members of groups (2), (3) and (4) (above) can be deficient in soils, due to low concentrations of the element in a particular parent rock or loss during weathering and leaching of rock and soil. Since a shortage of any one essential element can cause depressions in crop yield, even if all other factors are favourable, the supply of deficient nutrients by manures and fertilizers has become an important aspect of crop production.

Until recently, farmers supplied essential nutrients to crops either by applying manure and crop residues to the soil or by various land husbandry practices (rotations, fallowing, shifting cultivation, etc.). The moisture content and macronutrient (N, P and K) levels in farm manures can vary over an enormous range according to the species age, condition and diet of the livestock involved and also to the management systems employed. Thus there is a marked contrast between liquid pig slurry (typically < 5% dry matter, 0·4% N, 0·1% P and 0·2% K, expressed on a fresh weight basis) and poultry manure (typically 30% dry matter, 1·7% N, 0·6% P and 0·6% K). Clearly, their high moisture content and variability in composition, give rise to serious practical problems in the use of manures for accurate fertilization of crops. On the other hand, manures are particularly useful in supplying adequate and balanced quantities of micronutrients, although slurry from intensive livestock enterprises can contain very high, and potentially toxic, levels of Cu. Other organic materials which have been used widely in agriculture and horticulture include sewage sludge, blood, bone meal, fish scrap, green manures, guano (sea bird manure) and compost.

Due to a large number of factors (e.g. separation of arable from livestock farming, inadequate supplies of manure, variability in the quality of manures, high transport costs and the high nutrient requirements of improved crop varieties— in particular the newer dwarf cereals), synthetic, inorganic fertilizers are used widely in intensive agriculture, sometimes in conjunction with organic manures. This chapter examines some relevant chemical characteristics of these synthetic fertilizers, concentrating on N, P and K compounds.

15.1 NITROGEN FERTILIZERS

Although nitrogen is very abundant in the atmosphere (78%), it exists in the form of very stable and unreactive diatomic molecules. Consequently, the conversion of atmospheric nitrogen into those forms which are useful to non-leguminous plants (i.e. NH_4^+ and NO_3^-) involves industrial processes requiring considerable quantities of energy (high pressure and high temperature reactions—see sections 7.5 and 7.6). Before the development of the Haber Process for the synthesis of ammonia about 65 years ago, farmers depended mainly upon manures, nitrogen-fixing legumes (in various rotation systems), guano (seabird manure from S. America—up to 13% N and 9% P) and Chile Nitrate, for the supply of nitrogen to crops. Since then, and particularly since 1940, synthetic nitrogen fertilizers have become increasingly important, with annual production of fertilizer from the Haber Process alone exceeding 100 million tonnes.

Chemical Properties of Nitrogen Fertilizers (Table 15.1)

Ammonium sulphate, originally a by-product of coal–gas generation, was for a long time the most popular nitrogen fertilizer, but it has now been widely superceded by ammonium nitrate (the most important compound in the UK), urea and ammonia, because of their higher nitrogen contents. Anhydrous ammonia is used particularly

TABLE 15.1. Properties of nitrogen fertilizers.

Fertilizer	Chemical substance(s)	Solubility (g 100 ml^{-1} water)	% N	Preparation or source
Ammonium sulphate	$(NH_4)_2SO_4$	71	21	From NH_3 and H_2SO_4 (section 7.5)
Sodium nitrate	$NaNO_3$	92	16	Chile nitrate deposits
Ammonium nitrate	NH_4NO_3	118	35	From NH_3 and HNO_3 (section 7.5)
Nitrochalk*	$\begin{cases} NH_4NO_3 \\ \text{and} \\ CaCO_3 \end{cases}$	118 0·0014	20–26	From NH_3 and HNO_3 (section 7.5) with added limestone
Urea	$CO(NH_2)_2$	100	47	From NH_3 and CO_2, at high pressure and temperature
Anhydrous ammonia	NH_3	90	82	Haber Process (sections 7.5 and 7.6)

* Nitrochalk is the trade name for this product in the UK; also called Ammonium Nitrate with Lime (ANL) in the USA and Calcium Ammonium Nitrate (CAN) elsewhere.

on highly mechanized farms in North America where it is stored as a liquid under pressure and injected into the soil or into irrigation water.

Since all nitrogen fertilizer substances are highly soluble in water (section 5.2; Table 15.1), they are readily available in the soil solution for uptake by plant roots. However, their high solubilities can also result in substantial losses of nitrogen by leaching if heavy rainfall follows fertilizer application. Because of this and other losses (e.g. denitrification) only 40 to 60% of applied fertilizer nitrogen can be recovered in the grain and straw of cereal crops in the UK. Higher recovery rates, 60–75% can be obtained from grassland.

A common feature of ammonium-containing fertilizers is their tendency to lower the pH of soils. There are two reasons for this: firstly, ammonium ions bind to the soil colloids (by ion exchange reactions) and are gradually converted to nitrate by soil microbial activity:

$$NH_4^+ + 2O_2 + H_2O \rightarrow NO_3^- + 2H_3O^+ \text{ (nitrification)}.$$

thus increasing the hydronium ion concentration of the soil solution. This reaction also leads to the second effect: nitrate ions (from the fertilizer and from nitrification) tend to be leached out of the soil in drainage water. However, the fertilizer cations (NH_4^+) have disappeared during nitrification and therefore, in order to maintain electroneutrality, metal cations (especially K^+ and Ca^{2+}) must accompany the nitrate ions and are lost from the soil to the groundwater. Thus the natural tendency for well-drained soils to become more acidic with time (section 9.5), is intensified by the application of ammonium fertilizers. This effect can largely be avoided by the addition of ground limestone to the fertilizer, as in Nitrochalk (Table 15.1), or by using nitrate fertilizers only.

Although urea releases ammonium ions in contact with soil, by the reaction:

$$CO(NH_2)_2 + 2H_2O \rightarrow (NH_4)_2CO_3$$

this does not lead to acidification because of the low solubility of carbonate ions. However, care must be exercised when applying urea and ammonia to crops since high concentrations in soils can be phytotoxic.

Care must also be exercised in the storage and handling of nitrogen fertilizers; ammonium nitrate is particularly troublesome since it is hygroscopic (absorbs moisture from the atmosphere) and also a strong, and potentially explosive, oxidizing agent. Some fertilizers provide significant quantities of other essential nutrients. For example, ammonium sulphate contains 24% S, whereas Chile Nitrate (from natural mineral deposits in South America) provides varying quantities of trace elements.

15.2 PHOSPHORUS FERTILIZERS

Plants absorb phosphorus from the soil in the form of phosphate ions ($H_2PO_4^-$ or HPO_4^{2-}, see Figure 8.1). The earliest phosphorus fertilizers were ground-up bones or rock phosphate, both of which contain the rather insoluble phosphate mineral, apatite (Table 15.2). However, around 1840, it was discovered that treating bones or rock phosphate with sulphuric acid gave much more soluble phosphates which were, therefore, more readily available for plant uptake. The product from rock phosphate was called *superphosphate* (later becoming *single* or *ordinary* superphosphate) but this has largely been superceded by *concentrated*, *double* or *triple* superphosphate (different names for the same product) whose higher phosphorus content (Table 15.2) is due to the substitution of phosphoric acid for sulphuric acid. The use of bones as fertilizers is now restricted to horticultural applications of bone meal.

Chemical Properties of Phosphorus Fertilizers (Table 15.2)

In contrast to nitrogen fertilizer compounds, even the more soluble superphosphate fertilizers are only sparingly soluble in water. Consequently, the 'available' phosphorus in soils includes phosphate ions in the soil solution *and* those 'labile' solid phosphates which will dissolve readily to replace phosphate ions which have been absorbed by the crop. The aim of phosphorus fertilization is, therefore, to add to the soil, phosphorus compounds which will maintain soil solution concentrations of phosphate at levels ($10^{-4} - 10^{-5}$ M) which are adequate for rapid crop growth.

All phosphorus fertilizers contain a proportion of phosphate compounds which are not labile due to their very low solubility products (e.g. hydroxyapatite, Table 5.2). Since it is difficult to measure the solubility of phosphates *in soil*, fertilizer manufacturers attempt to predict the availability of fertilizer phosphorus by measuring how much will dissolve in certain aqueous solutions. For example, the available phosphate content of the more soluble fertilizers (superphosphates) is assumed to be the proportion of the phosphate content which is soluble in water, whereas for the less soluble materials (rock phosphate, basic slag) the test solvent is normally a two per cent citric acid solution. Typical values for commercial fertilizers are given in Table 15.2.

However, such expressions of availability are not very useful in practice since substances like monocalcium phosphate, $Ca(H_2PO_4)_2$, which are initially available in soils tend to react to give much more insoluble phosphates of iron and aluminium (e.g. Variscite, Table 5.2). Consequently, the availability of added phosphate depends upon the period of uptake and the chemical properties of the soil, and is best measured by the amount of phosphorus actually taken up by a crop under standard conditions. Arable crops can recover up to 25% of fertilizer phosphate (superphosphate) in the season of application and much smaller amounts in succeeding seasons.

Because of the low content and availability of phosphorus in rock phosphate and basic slag (Table 15.2), these are of limited use as arable fertilizers, in spite of their low cost. However, the phosphate in these materials is more soluble in acid, than neutral, soils and is

TABLE 15.2. Properties of phosphorus fertilizers.

Fertilizer	Chemical substance	Solubility (g 100 ml^{-1} water)	% P	% of P in fertilizer available to plants
*Rock phosphate	Apatite $Ca_{10}(PO_4, CO_3)_6(F, Cl, OH)_2$	sparingly soluble	6-18	14-65
Ordinary/single superphosphate	Monocalcium phosphate $Ca(H_2PO_4)_2$ + Gypsum $CaSO_4$	1·8 0·209	8-10	97-100
Concentrated/triple superphosphate	$Ca(H_2PO_4)_2$	1·8	19-23	96-99
†Basic slag	Complex mixture of Ca silicates and phosphates	variable (sparingly soluble)	3-10	62-94

* Obtained from natural deposits in North Africa, the USSR and the USA.
† By-product of the blast furnace—see section 4.3.

released at a slow rate over a number of years. Consequently, these materials, as fine powders, are useful in the nutrition of perennial crops, especially upland grassland. All phosphate fertilizers contribute considerable quantities of Ca to the soil, and basic slag is particularly useful in supplying varying amounts of S, Mg and trace elements.

Early investigators thought that P_2O_5 was the active substance in phosphorus fertilizers and so they expressed the total phosphorus content as % P_2O_5. Although we now know that phosphate ions are the chemical species taken up by roots, the habit of expressing P content as % P_2O_5 persists to the present. To interconvert % P_2O_5 and % P, we use the following expressions:

$$\% \ P \ \ \ \ = \ \% \ P_2O_5 \times 0\cdot44$$
$$\% \ P_2O_5 = \ \% \ P \ \ \ \ \times 2\cdot27.$$

TABLE 15.3. Properties of potassium fertilizers.

Fertilizer	Chemical substance	Solubility (g 100 ml^{-1})	% K
* Potassium chloride (muriate of potash)	KCl	35	52
Potassium sulphate	K_2SO_4	12	44
Potassium nitrate (saltpetre)	KNO_3	13	39

* Obtained from natural deposits of Silvite in Canada, Germany and the USSR.

15.3 POTASSIUM FERTILIZERS

Traditionally, shortage of potassium in soils was alleviated by the application of ashes of wood, crop residues or seaweed, and from this practice comes the name potash. Potassium is now generally supplied in the form of simple, soluble ionic salts, principally potassium chloride, although potassium sulphate or nitrate are used where a crop is sensitive to high chloride levels (e.g. tobacco). In spite of the high solubility of potassium chloride, leaching losses are modest due to the binding of K ions to soil colloids by ion exchange. In the season of application, cereals can absorb 50% and grassland up to 80% of fertilizer potassium from a fertile soil.

As for phosphorus fertilizers, the potassium content of a fertilizer is normally given as the percentage of the oxide, K_2O. To interconvert with % K, the following expressions are used:

$$\% \text{ K} \quad = \ \% \text{ K}_2\text{O} \times 0{\cdot}83$$
$$\% \text{ K}_2\text{O} = \ \% \text{ K} \quad \times 1{\cdot}2.$$

15.4 LIMING MATERIALS

Calcium ions must be supplied to agricultural soils for at least three reasons. For:
   (a) supply of the macronutrient Ca to plants;
   (b) maintenance of optimum soil pH (see sections 8.1 and 8.2);
   (c) maintenance of soil structure (aggregation—see section 10.5).
In intensively cultivated soils, these requirements are normally met by regular liming to the required pH. However, in addition to supplying other essential elements (N, P, S, etc.), some fertilizers also supply calcium as shown in Table 15.4.

TABLE 15.4. The calcium content of some fertilizers*.

Fertilizer	Supplying	Mean Ca content (%)
Nitrochalk	N	8
Gypsum	S	22
Rock phosphate	P	33
Single superphosphate	P	20
Triple superphosphate	P	14
Basic slag	P	32

* Note that these are not liming materials.

Chemical Properties of Liming Materials

Liming materials, which contain calcium ions in association with *non leachable* anions (CO_3, OH), include:
   Calcium carbonate, $CaCO_3$, normally as finely crushed limestone, but occasionally as chalk or marl. Dolomitic limestone, a mixed $CaCO_3/MgCO_3$ limestone, has the additional value of supplying Mg. All limestones can be stored and handled without difficulty.
   Calcium oxide, CaO (lime, quicklime, burnt lime, unslaked lime) is prepared by the thermal decomposition of limestone in lime kilns:

$$CaCO_3 \rightarrow CaO + CO_2\uparrow.$$

Quicklime, an unpleasant fine powder which burns the skin, absorbs moisture to give hard lumps of calcium hydroxide and carbonate.

Calcium hydroxide, $Ca(OH)_2$ (slaked lime, builder's lime, hydrated lime), which is prepared by the controlled slaking of quicklime:

$$CaO + H_2O \rightarrow Ca(OH)_2$$

is less caustic to the skin than quicklime.

Although CaO and $Ca(OH)_2$ are more reactive and act more quickly in the soil, ground limestone is the most commonly used agricultural lime due to its lower cost and its superior handling and storage properties.

15.5 COMPOUND FERTILIZERS

If two or more nutrients are required by the same crop, then it is normally more convenient and more economic to apply them together as a compound fertilizer. A great variety of different compound fertilizers have been manufactured to meet the requirements of different crops and soils, and their composition is normally expressed by the $\% \, N : \% \, P_2O_5 : \% \, K_2O$ ratio, more commonly written as the NPK ratio.

As a particular example, the ferruginous soils on the plateau areas of Malawi are deficient in phosphorus and nitrogen but rich in potassium, released by the weathering of hornblende minerals. Consequently, the most commonly used general-purpose fertilizer (especially for maize) is 20 : 20 : 0, i.e. $20\% \, N : 20\% P_2O_5 : O\% \, K_2O$. However, where cotton is grown on hydromorphic soils in the lower altitude areas, it is essential to restrict nitrogen supply to avoid excessive vegetative growth (plants taller than 2 m are not unusual). For this purpose it is customary to use 2 : 18 : 15, i.e. $2\% \, N : 18\% P_2O_5 : 15\% \, K_2O$.

This ratio gives only the quantities of nutrients present in the fertilizer and reveals nothing about the chemical nature of the ingredients. The 'recipe' for a particular compound fertilizer may vary widely according to:

(a) the *compatibility* of chemicals mixed together, e.g. an acid compound will not be compatible with a basic compound;

(b) the relative *costs* of different compounds supplying the same element;

(c) the *commercial availability* of fertilizer compounds.

Most compound fertilizers and many 'straight' (single nutrient) fertilizers are now produced in granular form. Granulation reduces the caking of fertilizer particles, encourages the free running of particles in fertilizer spreaders and cuts down dust hazards. In most compound fertilizers, each granule contains the ratio of nutrients printed on the bag, thus ensuring uniformity of nutrient distribution.

FURTHER READING

1. BRADY N.C. (1974) *The Nature and Properties of Soils*, 8th Edition, Ch. 2. 15–20. Macmillan, New York.
2. EPSTEIN E. (1972) *Mineral Nutrition of Plants: Principles and Perspectives*, Part 1. John Wiley & Son, New York.
3. TISDALE S.L. & NELSON W.R. (1975) *Soil Fertility and Fertilisers*. Collier Macmillan, New York.
4. WHITE R.E. (1979) *Introduction to the Principles and Practice of Soil Science*, Ch. 12. Blackwell Scientific Publications, Oxford.
5. *Chemistry and Agriculture* (1979) Special Publication 36, The Chemical Society.

CHAPTER 16
PESTICIDES

Pesticides are chemical substances used to kill organisms (bacteria, fungi, insects, nematodes, rats, birds, etc.) which consume, damage or destroy agricultural products, either in the field or during storage. The term 'pest' is also extended to include weeds, which, by competing with crop plants, reduce yields. Most, but not all, modern pesticides are complex organic substances whose detailed chemistry is beyond the scope of this book. However, by studying the composition, solubility and overall reactivity of pesticides, it is possible to classify them according to their activity and persistence in the environment.

Because of their chemical complexity, pesticide substances tend to have long and cumbersome systematic names. Consequently, manufacturers tend to use simple, more easily remembered, non-systematic names both for the pesticide substance and the commercial product. This can lead to a degree of confusion, since a pesticide can have three names—a product or trade name, a non-systematic name for the active ingredient and a systematic chemical name. To avoid such confusion, commercial product names have been omitted from the following sections.

16.1 INSECTICIDES FOR CROP PROTECTION

The ideal insecticide would be a chemical substance whose action was specific to the target pest (i.e. non-toxic to other arthropods), which was harmless to all other members of the ecosystem, including the pesticide operator, and which was degraded to a harmless residue whenever its action was complete. The development of such a model pesticide is still a very long way away, but the evolution of insecticides over the past 50 years has been guided to a considerable extent by these ideals. For example, before the development of the first synthetic insecticides about forty years ago, insect and other arthropod pests were generally treated with a variety of inorganic compounds containing copper, lead, arsenic, fluorine or boron:

e.g.

Paris Green, $Cu_4(CH_3CO_2)_2(AsO_2)_2$;
Lead arsenate, $PbHAsO_4$;
Sodium fluoride, NaF;
Sodium silicofluoride, Na_3SiF_6;
Borax, $Na_2B_4O_7 \cdot 10H_2O$;
Cryolite (mineral), Na_3AlF_6,

which were far from being ideal (non-specific, high mammalian toxicity, non-degradable and very persistent in soils). These substances have subsequently been replaced by a variety of organic pesticides which, after serious problems with both organochlorine and organophosphorus insecticides, are much closer to the ideal. The various classes of organic insecticides are as follows:

(a) Naturally Occurring Organic Insecticides

During the 18th and 19th centuries it was discovered that a variety of plant extracts used for other purposes (e.g. poisoning fish), were useful insecticides. The most important of these, commercially, are as follows.

(i) Nicotine, $C_{10}H_{14}N_2$—

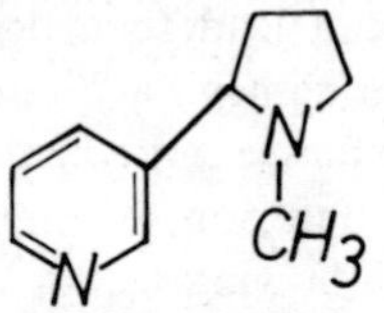

an alkaloid (section 14.1) from tobacco, commonly used in the fumigation of greenhouses.

(ii) *Pyrethrum* extracts from the flowers of *Pyrethrum cinerariaefolium* grown in East Africa. There are several active ingredients including *Pyrethrin* 1, $C_{21}H_{28}O_3$—

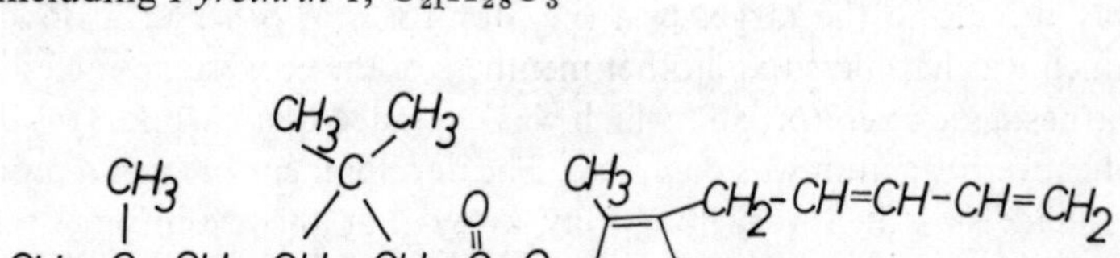

Due to its rapid 'knockdown' activity and low toxicity to mammals, pyrethrum is an important insecticide, for example, in household sprays and in horticulture. The recent laboratory synthesis of new pyrethrin molecules may lead to increased use of these substances in agriculture.

(iii) Extracts from several *Derris* species from South America, whose active ingredient is *Rotenone*, $C_{23}H_{22}O_6$—

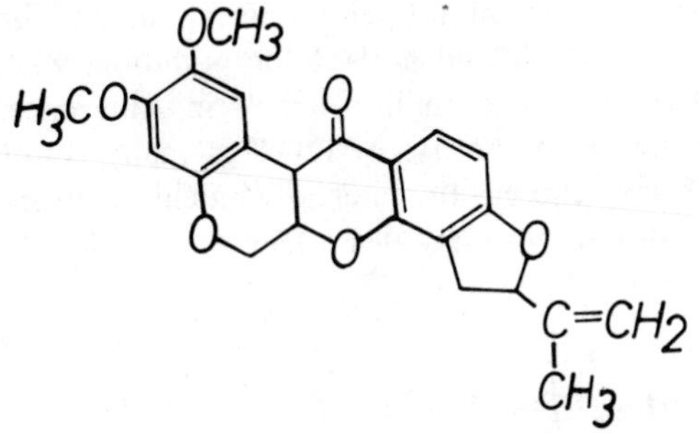

are also mainly employed in horticulture, where low mammalian toxicity is extremely important. Nicotine, pyrethrin and derris are also non-persistent, breaking down to harmless residues within a few hours of application.

(iv) In addition to these natural organic pesticides, various hydrocarbon oils (aliphatic and aromatic) have been used to kill insects and mites and continue to be used in certain cases, especially in fruit tree orchards.

(b) Organochlorine Insecticides

This group of synthetic insecticides includes a variety of aromatic and non-aromatic ring compounds, all containing chlorine. Due to the aromatic nature of some of these compounds and the lack of hydrophilic groups in all members, these pesticides are highly hydrophobic and very resistant to chemical degradation. The first successful member of this group, introduced in 1939/40 was DDT ($C_{14}H_9Cl_5$, 1,1,1-trichloro-2,2-di-(4–chlorophenyl) ethane—

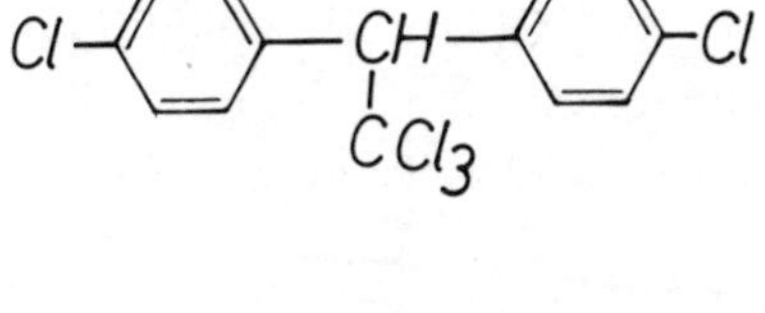

which, because of its low mammalian acute toxicity, was used widely and in large quantities for public health applications (against human ectoparasites, malaria mosquitoes, etc.) and subsequently in agriculture, horticulture and forestry.

More recently, it was discovered that due to its unreactivity and hydrophobic properties, DDT and its residues are extremely persistent in the environment and tend to accumulate in the lipid stores of long-lived mammals and birds. Such pollution, which is evident throughout the world, even in seabirds in Antarctica, has led to stringent controls on the use of DDT in many countries.

The problems caused by later organochlorine pesticides were much more serious. For example, Dieldrin ($C_{12}H_8Cl_6O$) (I),

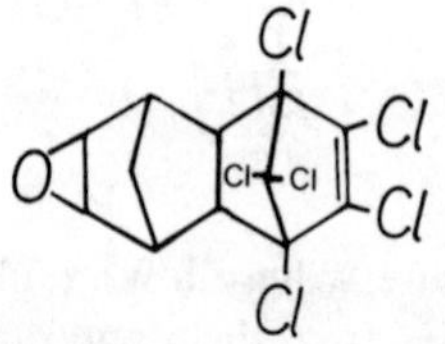

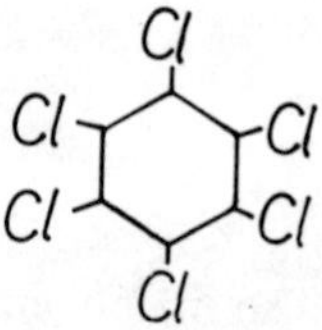

used as a seed treatment and sheep-dip, caused widespread mortality amongst birds eating newly sown grain, and in predatory birds, due to the persistence of the pesticide. Use of the more toxic products, dieldrin, aldrin, heptachlor, endosulphan and endrin is now severely restricted in most countries, but Lindane or γ-BHC ($C_6H_6Cl_6$ 1,2,3,4,5,6-hexachloro-cyclohexane) (II) is still used for many applications due to its lower mammalian toxicity and more rapid breakdown in soils and tissues.

(c) Organophosphorus Insecticides

This group of synthetic insecticides includes a vast array of aliphatic and aromatic compounds containing phosphorus, which, like the organochlorines, kill insects by chemical disruption of the nervous system. All have the general formula:

$$G_1 - \overset{\overset{\textstyle O \text{ (or S)}}{\|}}{\underset{\underset{\textstyle G_2}{|}}{P}} - X$$

where the G_1 and G_2 groups are attached strongly and stably to the P atom, whereas the X group is more easily removed by hydrolysis. These insecticides are, therefore, less stable chemically than the organochlorines and tend to be degraded to harmless residues within hours of application. However, their reactivity also makes organophosphorus compounds harmful to mammals by interfering with enzyme function. Thus this group of pesticides can cause acute toxicity and death (of humans, livestock and wildlife) at the time of application but they do not lead to the accumulation of potentially harmful residues in the environment. They are, therefore, more 'ideal' insecticides, if applied carefully and safely. (Note that the water solubility of organophosphorus pesticides varies considerably according to the nature of G_1, G_2 and X).

Parathion ($C_{10}H_{14}NO_5PS$)—

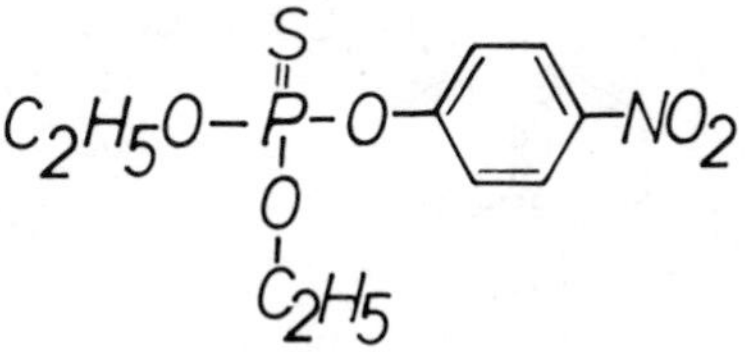

$$C_2H_5O-\underset{\underset{C_2H_5}{\overset{|}{O}}}{\overset{\overset{S}{\|}}{P}}-O-\langle\overline{}\rangle-NO_2$$

was the first organophosphorus insecticide to be used in agriculture (from 1944), but due to its high human toxicity it has been generally replaced by a large number of less toxic, and more specific, compounds such as:

Phosphamidon $\quad G_1, G_2 = OCH_3$

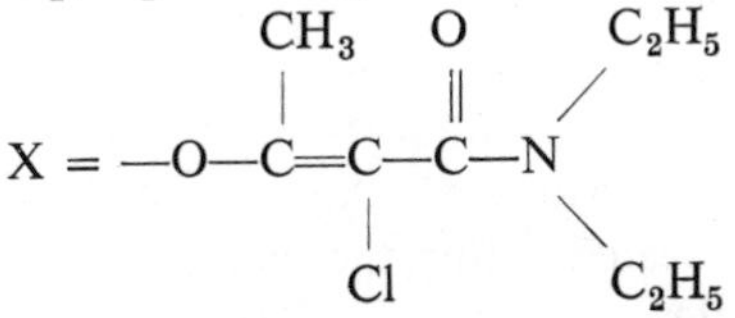

$$X = -O-\underset{\underset{Cl}{\overset{|}{}}}{\overset{\overset{CH_3}{\overset{|}{}}}{C}}=C-\overset{\overset{O}{\|}}{C}-N\overset{\diagup C_2H_5}{\diagdown C_2H_5}$$

Dimethoate $\quad G_1, G_2 = OCH_3$

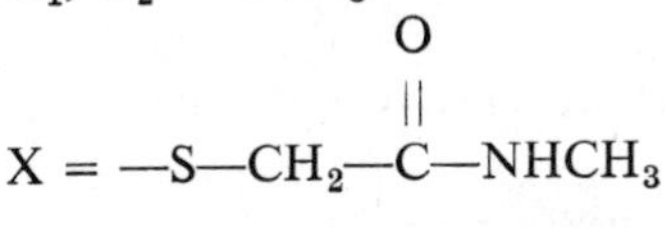

$$X = -S-CH_2-\overset{\overset{O}{\|}}{C}-NHCH_3$$

(d) Carbamate Insecticides

This third main group of synthetic agricultural insecticides includes a variety of compounds of general formula:

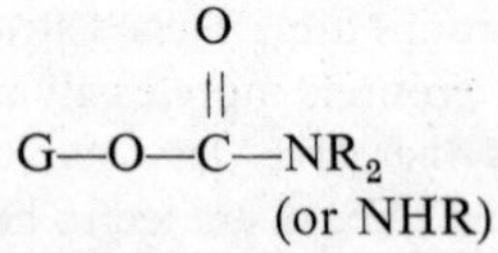

$$
\begin{array}{c}
O \\
\parallel \\
G\!-\!O\!-\!C\!-\!NR_2 \\
(or\ NHR)
\end{array}
$$

where G is normally aromatic or heterocyclic, and R is almost invariably a methyl group. Carbamates are generally similar to organophosphorus compounds in their mode of action and persistence, and vary considerably in their water solubility according to the nature of the G group (e.g. Carbaryl 120 mg l^{-1}, Aldicarb 6 g l^{-1}). The most important carbamate insecticide in use is Carbaryl $C_{12}H_{11}NO_2$, 1-naphthyl methylcarbamate)—

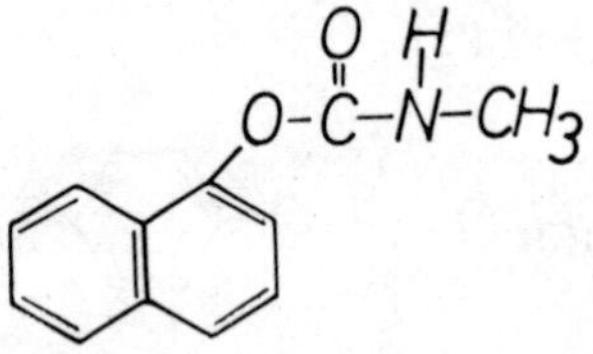

which combines low mammalian toxicity with rapid degradation after application. Other useful carbamates include Aldicarb $(C_7H_{14}N_2O_2S)$:

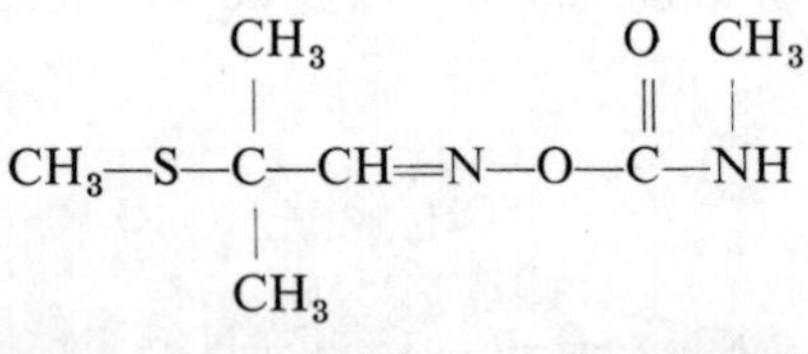

which, like several organophosphorus pesticides, is systemic (absorbed and translocated within the plant, giving protection against sap-feeding insects).

16.2 FUNGICIDES FOR CROP PROTECTION

Inorganic compounds have been used for a long time to protect the surfaces of crop plants against fungal attack. For example, elementary sulphur has been used since the eighteenth century, if not earlier, and insoluble copper compounds first appeared in 1885 with the development of Bordeaux mixture, a sticky paste of $Ca(OH)_2$ and $CuSO_4$

$5H_2O$, for use on grape vines. Because of the greater susceptibility to copper of fungi (compared with higher plants), numerous copper-containing substances were developed subsequently, but the main compound still in use for foliar protection is copper oxychloride, $Cu_2(OH)_3Cl$, which is virtually insoluble in water but does form colloidal suspensions. Inorganic compounds of Zn, Cr, Ni, Hg, Sn and Pb have also been tried without outstanding success.

Unlike insecticides, synthetic fungicides cannot be divided into a few major chemical groups because of the great variety of substances involved—organometal compounds, carbamates, organophosphorus compounds, quinones, nitro compounds, various heterocyclic and aromatic compounds. Instead, fungicides can be classed according to their function, i.e.

(a) *Foliar (Protectant) Fungicides* which prevent the spread of fungal infection, presumably by killing spores. This role is filled principally by copper compounds as described above but more recently developed synthetic pesticides such as Maneb ($C_4H_6MnN_2S_4$)

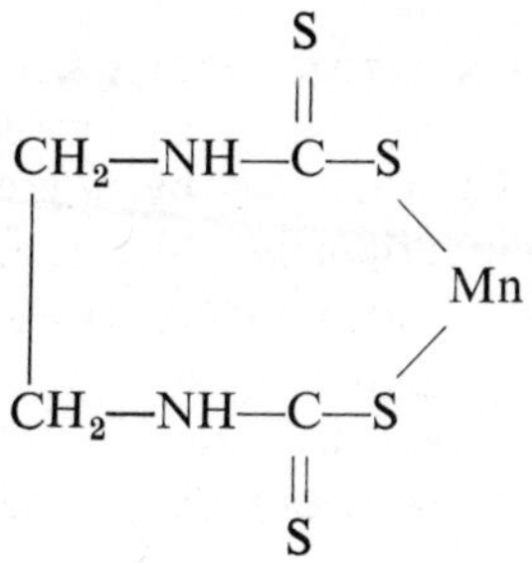

$$
\begin{array}{c}
\overset{\displaystyle S}{\overset{\displaystyle \|}{}} \\
CH_2-NH-C-S \\
| \qquad\qquad\qquad \searrow \\
\qquad\qquad\qquad\qquad Mn \\
| \qquad\qquad\qquad \nearrow \\
CH_2-NH-C-S \\
\overset{\displaystyle \|}{\underset{\displaystyle S}{}}
\end{array}
$$

(note that Zineb is identical but with a Zn atom substituted for the Mn) and Chlorothalonil ($C_8Cl_4N_2$; 2,4,5,6-tetrachloro-1, 3-dicyanobenzene)—

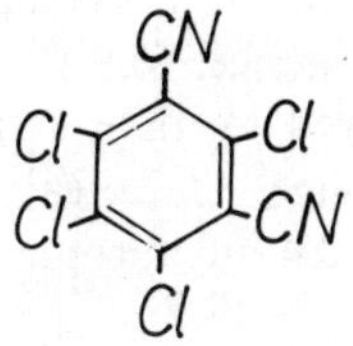

are widely used, frequently in association with inorganic compounds.

(b) *Systemic Fungicides* which are absorbed by the plant and distributed throughout its tissues giving protection against a wide range of fungal pathogens including leaf-infecting organisms. This class includes a range of relatively soluble heterocyclic compounds including Benomyl ($C_{14}H_{18}O_3N_4$)—

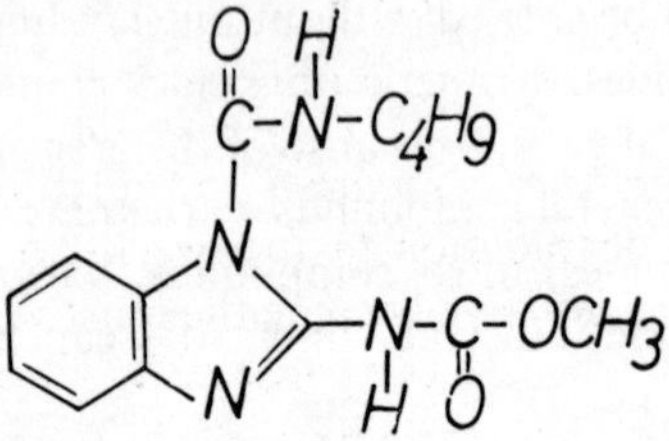

(c) *Fungicides for Seed Treatment* against soil pathogens and seed-borne diseases, include organomercury compounds as well as several pesticides classed as systemic.

16.3 WEEDKILLERS/HERBICIDES

Although weeds can be destroyed by a great variety of strong chemicals including sulphuric acid, chlorates, borates, copper salts and various organic oils (creosote, etc.), these substances tend to kill all living plants and cannot be used to remove unwanted weeds growing amongst crop plants. Three important classes of more selective weedkillers, which are essential for mechanized agriculture, are discussed below.

(a) *Plant Growth Substances as Weedkillers* (Phenoxyacetic acid derivatives). About 40 years ago, it was discovered that certain synthetic organic substances affect the growth of plants in a way similar to auxin (section 14.1). The application of relatively large quantities of these substances to weeds causes serious disorganization in growth followed by the death of (dicotyledonous) weed plants, without damage to cereal crops. Examples include: 2,4-D ($C_8H_6Cl_2O_3$, 2,4-dichlorophenoxyacetic acid—solubility $0{\cdot}6$ g l^{-1})—

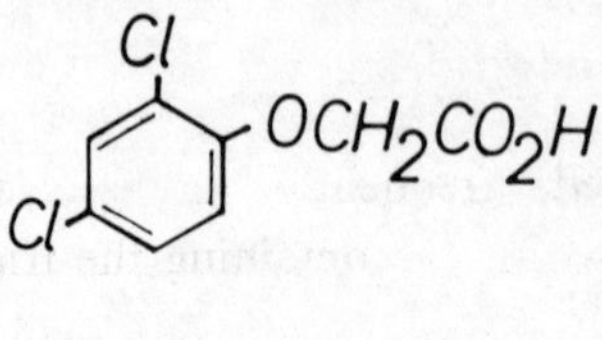

MCPA ($C_9H_9ClO_3$, 4-chloro-2-methylphenoxyacetic acid—solubility $0 \cdot 8$ g l^{-1})—

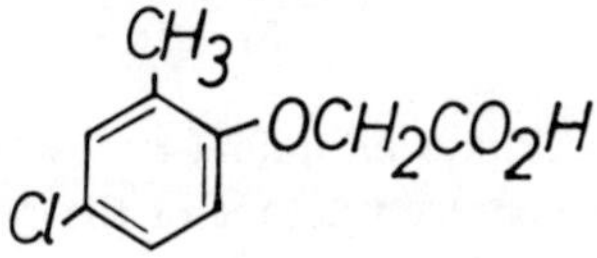

$$CH_3$$
$$Cl\text{—}\text{—}OCH_2CO_2H$$

Because of their high solubility, low mammalian toxicity and relatively rapid degradation (a few weeks in soil), the phenoxyacetic acid herbicides have become the most widely used pesticides in the UK. However, concentration on killing the most troublesome dicotyledonous weeds has resulted in an upsurge in perennial grass weeds. To meet this problem,

(b) *The Bipyridylium Herbicides* were developed. For example, the most important member, Paraquat ($C_{12}H_{14}N_2^{2+}$, $1,1^1$-dimethyl-4, 4^1-bipyridylium ion)—

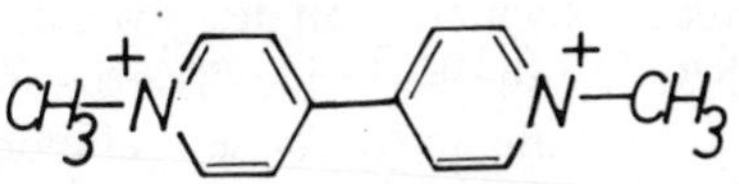

$$CH_3\overset{+}{\text{—}N}\text{—}\text{—}\overset{+}{N}\text{—}CH_3$$

which exists as a quaternary ammonium ion (section 14.1), is very effective at killing monocotyledonous weeds when applied between successive crops and in preparation for direct drilling. It can persist for many years in soils but in an inert, non toxic form bound to clay particles. The bipyridylium herbicides are now to a certain extent being superceded by a relatively simple aliphatic chemical, glyphosate:

$$\overset{\displaystyle O}{\underset{\displaystyle OH}{\overset{\displaystyle \|}{HO\text{—}P}\underset{\displaystyle |}{}}}\text{—}CH_2\text{—}NH\text{—}CH_2\text{—}CO_2H$$

which is absorbed by grass leaves and translocated into underground rhizomes. It is therefore very useful in dealing with weeds such as *Agropyron repens*.

(c) *The Triazines* are another interesting group of selective herbicide compounds, all containing the triazine ring structure—

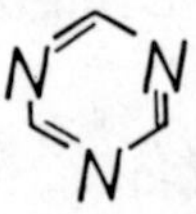

used to kill the seedlings of annual weeds in crops, such as maize and soft fruit, which are tolerant of the pesticide. For example, Simazine $(C_7H_{12}ClN_5)$—

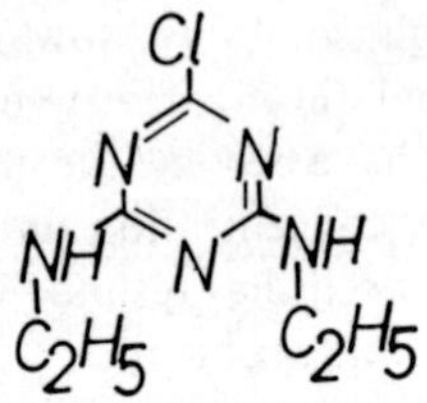

applied to the soil before the emergence of a maize crop will, in spite of its low solubility $(3 \cdot 5 \text{ mg l}^{-1})$, be taken up into the seedlings of emerging weeds which are then destroyed due to the disruption of photosynthesis by the herbicide. Triazines are not subject to leaching and can remain active for several months, bound to soil colloids.

16.4 OTHER AGRICULTURAL PESTICIDES

In many areas, it is essential to dip livestock in order to remove or repel disease-carrying arthropods such as ticks. Dip chemicals which are widely used include arsenic trioxide (highly soluble and toxic) as well as more conventional insecticides such as camphechlor (organochlorine), dioxathion and chlorfenvinphos (organophosphorus). Several commercial formulations contain combinations of pesticides, e.g. camphechlor + dioxathion. The disposal of spent dip solutions poses serious environmental problems (section 9.4).

Other applications of pesticides include the fumigation of tobacco nursery and seed potato soils (to kill nematodes) and of grain stores (to kill insects). Fumigant substances in use include formaldehyde (HCHO), carbon disulphide (CS_2), chloropicrin (Cl_3CNO_2), methyl bromide (CH_3Br) and ethylene dibromide or EDB $(CH_2Br—CH_2Br)$. Other pesticides are used to deal with snails and slugs (section 13.4), rats, birds and so on.

FURTHER READING

1. GREEN M.B., HARTLEY G.S. & WEST T.F. (1977) *Chemicals for Crop Protection and Pest Control.* Pergamon Press, Oxford.
2. MARTIN H. (1973) *The Scientific Principles of Crop Protection.* Edward Arnold, London.
3. MELLANBY K. (1967) *Pesticides and Pollution.* Wm. Collins Sons & Co. Ltd., Glasgow.
4. WHEELER B.E.J. (1976) *Diseases in Crops.* Studies in Biology No. 64. Edward Arnold, London.
5. WHITE R.E. (1979) *Introduction to the Principles and Practice of Soil Science,* Ch. 12. Blackwell Scientific Publications, Oxford.
6. WORTHING C.R. (1979) *The Pesticide Manual,* 6th Edition. British Crop Protection Council.
7. *Chemistry and Agriculture* (1979) The Chemical Society.

APPENDIX
EXERCISE SOLUTIONS

Chapter 1

1 B (5 protons, 6 neutrons, 5 electrons); O (8, 8, 8); Mg (12, 12, 12); Si (14, 14, 14); S (16, 16, 16); Ca (20, 20, 20); Co (27, 32, 27); Zn (30, 35, 30); As (33, 42, 33); I (53, 74, 53).

Chapter 3

	Liquid range (°C)	Density (g ml^{-1})	Surface properties
Ethanol	-117–79	0·79	hydrophilic
Carbon tetrachloride	-23–77	1·59	hydrophobic
Benzene	6–80	0·88	hydrophobic
Iodine	114–184	4·93	—

From this table it is clear that none of the four substances would be a complete substitute for mercury. Benzene and iodine are ruled out by the unsuitability of their liquid ranges whereas ethanol and carbon tetrachloride, although useful at low temperatures, could not be used above 75°C. In spite of its hydrophilic properties, ethanol (dyed for visibility) *has* been used widely in low temperature laboratory and meteorological thermometers, in which the problems due to the low density of the liquid are overcome by using glass tubes of very fine bore. Ethanol is preferred to other liquids such as carbon tetrachloride because of its low cost.

Chapter 5

	Solvent	Solute	Examples
1	Gas	Gas	O_2 in N_2 (air)
		Liquid	H_2O in air
		Solid	dust (finer than colloidal dimensions. section 10.4) in air
	Liquid	Gas	O_2 in H_2O
		Liquid	Ethanol in H_2O
		Solid	NaCl in H_2O
	Solid	Gas	gas trapped in solidified (molten) metal
		Liquid	amalgams (metal and mercury)
		Solid	alloys

2 (a) 58·5; 16; 132; 46; 265.
  (b) 117 g; 90 g; 0·6 g; 12 g.
  (c) 1·95 g; 0·0117 g.
3 (a) 7·95 g.
  (b) 12·4 g.
4 1·2 atm; 4·8 atm; 9·6 atm; 48 atm.
  osmotic pressure of 1 M urea = 24 atm.
  Therefore all of the solutions are hypotonic compared with the urea solution except the NH_4NO_3 solution which is hypertonic.
5 The fluids (blood, lymph, etc.) bathing animal cells are isotonic with the cell contents.

Chapter 6

1 10·2 g; 7·9 g; 3·2 g.

Chapter 7

2 Spontaneous reactions are always exergonic reactions which do not require activation energy. If energy were required to begin a reaction, it could not be spontaneous.
3 Increased temperature → Increased yield of HI.
  Increased pressure   → No effect (2 molecules produced by the reaction of 2 molecules, therefore no change in pressure).
  Product removal   → Increased yield of HI.

Chapter 8

1 3N; 3·4N; 0·1N; 0·018N.
2 400 ml; 16·5 ml; 210 ml.
3 If hydronium ions (i.e. a strong acid) are added to a solution of acetic acid at equilibrium:
$$CH_3CO_2H + H_2O \rightleftharpoons CH_3CO_2^- + H_3O^+$$
then, by Le Chatelier's Principle, the equilibrium position will move 'to the left' in order to remove hydronium ions.
4 As noted in section 8.4, for a buffer to work, both undissociated acid and anion must be present together in the solution. The undissociated acid is virtually absent in aqueous solutions of strong acids.

Chapter 9

1 The temperature of the surface of the Earth is determined primarily by its distance from the Sun. This temperature is at present about 15°C, on average, thus ensuring that most of the water in the hydrosphere is in the liquid state. Since all forms of life which have evolved on Earth are dependent upon the unique properties of *liquid* water (sections 9.2 and 9.3), a substantial movement of the Earth away from the Sun would result in the eventual extinction of all life due to the freezing of cells, as well as the water in the hydrosphere. A substantial movement towards the Sun would be equally disastrous due to the evaporation of the water present on Earth.
2 Liquid range of ammonia −78 to −33°C.

Chapter 10

1 The hydrophobic nature of liquid mercury causes it to be repelled from hydrophilic glass surfaces, resulting in a hemispherical convex meniscus.
2 The colloidal lipid globules in milk are stabilized by an amphipathic lipoprotein membrane (hydrophobic on its inner side and hydrophilic outside). During the churning of milk, this membrane is broken up by the action of air bubbles and by mechanical damage, permitting the lipid to flocculate to give butter.

Chapter 11

1 Familiar examples of elements with a valency of 4 include Si (in silicates, section 4.4) and N (in the ammonium ion, section 2.6). (In more complex compounds, elements like P and S can have valencies of 5 or 6.)

2 Total number of possible compounds = 70 (number of different combinations of 5 different things, 4 at a time, with repetitions—see any standard mathematical text, permutations and combinations).

Chapter 12

1 Pentane —n-pentane; 2-methyl-butane; 2,2-dimethyl-propane.

Hexane —n-hexane; 2-methyl-pentane; 3-methyl-pentane; 2,2-dimethyl-butane; 2,3-dimethyl-butane.

Heptane—n-heptane; 2-methyl-hexane; 3-methyl-hexane; 2,2-dimethyl-pentane; 3,3-dimethyl-pentane; 2,3-dimethyl-pentane; 2,4-dimethyl-pentane; 3-ethyl-pentane; 2,2,3-trimethyl-butane.

4 2 straight-chain pentenes —1-pentene; 2-pentene.

3 straight-chain hexenes —1-hexene; 2-hexene; 3-hexene.

Chapter 13

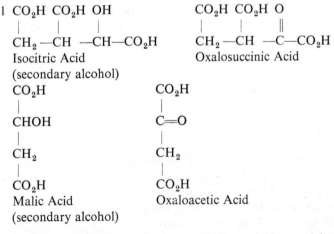

1
$$\begin{matrix} CO_2H & CO_2H & OH \\ | & | & | \\ CH_2 & —CH & —CH—CO_2H \end{matrix}$$
Isocitric Acid
(secondary alcohol)

$$\begin{matrix} CO_2H & CO_2H & O \\ | & | & \| \\ CH_2 & —CH & —C—CO_2H \end{matrix}$$
Oxalosuccinic Acid

$$\begin{matrix} CO_2H \\ | \\ CHOH \\ | \\ CH_2 \\ | \\ CO_2H \end{matrix}$$
Malic Acid
(secondary alcohol)

$$\begin{matrix} CO_2H \\ | \\ C=O \\ | \\ CH_2 \\ | \\ CO_2H \end{matrix}$$
Oxaloacetic Acid

2 1-methyl-4-hydroxy-benzene; 1,3-dihydroxy-benzene; 1-hydroxy-naphthalene; 1,2,3-trihydroxy-benzene; 1,3,5-trihydroxy-benzene.

3 (a) No; (b) No; (c) Yes; (d) Yes.

4 *Fehling's Solution*—alkaline solution of copper tartrate (blue solution containing Cu^{2+} ions)

$$
\underset{\substack{\text{aldehyde} \\ \text{blue} \\ \text{(oxidized)}}}{R\!-\!\overset{\displaystyle O}{\overset{\|}{C}}\!-\!H} + Cu^{2+} \rightarrow \underset{\substack{\text{acid} \\ \text{red} \\ \text{(reduced i.e. } Cu^+)}}{R\!-\!\overset{\displaystyle O}{\overset{\|}{C}}\!-\!OH} + Cu_2O\!\downarrow \qquad \text{(EQUATION NOT BALANCED)}
$$

ketones do not react.

The reaction of an aldehyde with Fehling's solution results in the loss of the blue colour and the production of a red precipitate of cuprous oxide. There are no colour changes with ketones.

Tollen's Reagent—ammoniacal silver nitrate solution.

$$
\underset{\substack{\text{aldehyde (oxidized)}}}{R\!-\!\overset{\displaystyle O}{\overset{\|}{C}}\!-\!H} + Ag^+ \rightarrow \underset{\substack{\text{acid} \quad \text{(reduced)}}}{R\!-\!\overset{\displaystyle O}{\overset{\|}{C}}\!-\!OH} + Ag\!\downarrow \qquad \text{(EQUATION NOT BALANCED)}
$$

ketones do not react.

The reaction of an aldehyde with Tollen's Reagent gives a precipitate of silver metal in the form of a 'silver mirror' on the inside of the test tube. Ketones do not give a silver mirror.

5 In maltose, the pyranose ring of one of the glucose units (i.e. the ring attached via C4) can open readily to give a reactive aldehyde group. It is, therefore, a reducing sugar. In contrast, the aldehyde group of the glucose unit in sucrose is involved in the glycosidic linkage between the two units and can react only after this linkage has been hydrolysed. Sucrose is, therefore, non-reducing.

6 $\dfrac{5\,000\,000}{180} = 27\,778.$

Chapter 14

1 The 'electron attracting power' of the benzene ring in aromatic amines, means that the nitrogen atom is less able to accept protons (see sections 2.6 and 13.2 and Chapter 8).

2 (a) All amino acids give a purple colour.

Imino acids (Proline and Hydroxyproline) give yellow colours.

(b) Violet or pink colours for proteins, peptides and urea only.

(c) Red colour with Arginine.

(d) (Hopkins–Cole Reaction). Purple ring with Tryptophan.
(e) Brown or black colours with Cysteine/Cystine.
(f) Yellow colours with amino acids containing a benzene ring, i.e. Phenylalanine, Tryptophan and Tyrosine.

REFERENCE

DIAMOND P.S. & DENMAN R.F. (1966) *Laboratory Techniques in Chemistry and Biochemistry.* Butterworth.

INDEX